James Freeman
Harris Twp. Fire Dept.

Fire Investigation

OTHER BOOKS BY THE AUTHOR

Crime Investigation: Physical Evidence and the Police Laboratory (Wiley, 1953)

Quantitive Ultramicroanalysis (Wiley, 1950)

Density and Refractive Index (Thomas, 1951)

The Crime Laboratory with L. W. Bradford (Thomas, 1965)

Fire Investigation

INCLUDING FIRE-RELATED PHENOMENA: ARSON, EXPLOSION, ASPHYXIATION

Paul L. Kirk, Ph.D.

Professor of Criminalistics, Emeritus
School of Criminology
University of California, Berkeley, Calif.

Paul L. Kirk, Ph.D., and Associates

Berkeley, California

John Wiley & Sons, Inc.,
New York, London, Sydney, Toronto

Copyright © 1969 by John Wiley & Sons, Inc. All rights reserved. No part of this book may be reproduced by any means, nor transmitted, nor translated into a machine language without the written permission of the publisher. Library of Congress Catalog Card Number: 69-19240 SBN 471 48860 7. Printed in the United States of America.

6 7 8 9 10

Table of Contents

1 INTRODUCTION   1

2 ELEMENTARY CHEMISTRY OF COMBUSTION   4

3 NATURE AND BEHAVIOR OF FIRE   14

4 COMBUSTION PROPERTIES OF NON-SOLID FUELS   28

5 COMBUSTION PROPERTIES OF SOLID FUELS   46

6 ROLE OF PYROLYSIS   63

7 FIRE PATTERNS OF STRUCTURAL FIRES   71

8 FIRE PATTERNS OF OUTDOOR FIRES   82

- 9 SOURCES OF IGNITION 89
- 10 AUTOMOBILE AND BOAT FIRES 119
- 11 CLOTHING AND FABRIC FIRES 125
- 12 PRACTICAL INVESTIGATION OF STRUCTURAL FIRES 134
- 13 ARSON 158
- 14 THE LEGAL ASPECT OF ARSON 173
- 15 CARBON MONOXIDE ASPHYXIATION 181
- 16 EXPLOSIONS ASSOCIATED WITH FIRES 193
- 17 BUILDING CONSTRUCTION MATERIALS 206

Appendices

- 1 FIRE EXPERIMENTATION 214
- 2 ILLUSTRATIONS OF FIRE ORIGINS 235
-   INDEX 251

Preface

For many centuries, fire has been one of the best friends of mankind. Its discovery was unquestionably one of the greatest boons, not excluding the discovery of the wheel or the later discovery of electricity. After all, there might be no electricity if fire were unknown, and there would be limited power to turn the wheel. It seems fair to state that modern civilization rests more directly on the knowledge and control of fire than on any other single item of human know-how. Not only does fire allow us to compete with a cold, inhospitable environment by providing heat for us, it also drives automobiles, trains, airplanes, and steamships. Atomic energy, generated without combustion, may ultimately replace some of its uses, as hydroelectric power has already done. Nevertheless, it seems fair to assume that even this potential boon to mankind will never replace fire completely. Unless the oil and coal reserves of the world are depleted, and the world population becomes so dense that it will be too expensive to maintain forests to produce wood, it is unlikely that man will ever outgrow his need for fire.

The contrast of fire, the friendly combustion, with fire, the destroyer, is striking. When fire turns from being a comforter and friend of man into a raging destructive enemy, a dilemma arises for which a solution is difficult to find. Even if man should replace all of his present need for the heat of combustion with heat from the atom, the tides, or the solar battery, it is doubtful if he can ever put aside the fear and loss engendered by an uncontrolled fire. As long as there are combustibles on earth, they will sometimes undergo combustion under circumstances that can be dangerous, frightening, and expensive. It is for this reason that the investigation of the causes of fire and their elimination will continue to be major preoccupations of a significant portion of earth's population.

It is scarcely predictable that fire insurance will ever become obsolete, or that the presently defined crime of arson will be taken off the books until the day when crime is defined as disease only. Even then, the fact of arson will still need to be established so that the actual perpetrator may be located and treated. Unless means can be found to induce honesty and morality in humans, it seems highly unlikely that the day will come when no building is destroyed by fire for monetary gain. In other words, there will be, for the foreseeable future, a real and pressing need for competent fire investigators.

It has long been noted that the excellence of investigators is largely conditioned by the demands of the legal profession. The inept, inexperienced, or unintelligent witness in the courtroom does not reappear often. The alert, capable, and reliable witness returns time and again. Only the legal profession has the opportunity of choice and of discrimination. Thus, the practitioners of law, more than any other group in our system, must be informed because that group ultimately will set the standards by which their investigators must operate. Without exposure to the facts and to the current practical knowledge, the legal profession is in no position to make the important discriminations and to demand the type of service that is its due. The competitive desire to win cases is, on the whole, the most important single factor upgrading the performance of all who work with attorneys. Certainly no investigator is closer to the attorney than the investigator of a fire that is of great criminal importance and carries with it a huge economic price tag. The same may be said of accident investigators. However, many more useful publications and practical studies are available to them than to investigators concerned with fires. It is hoped that this volume can serve to guide the attorney in his choice of competent fire investigators and also give him some information as to what constitutes high-grade performance.

Search of the literature reveals that there has been a large number of highly competent and illuminating publications on various aspects of fires and explosions. There is no dearth of data and facts. However, the most significant publications are in highly technical journals. While they are completely understandable to the trained chemist, they would rarely make sense to the average fire investigator or attorney. Seldom, indeed, does the professional chemist have an opportunity or need to investigate a fire; if he had, his ordinary total lack of investigative training and experience would make him a very poor fire investigator.

Many pamphlets and brochures are available to fire investigators, but most of these show a preoccupation with issues that are extraneous to the fire investigation itself. The theme may be legal, may concern the types of persons who become firebugs and how to recognize them, or may have to do with statistical considerations that are of interest from the legal or economic standpoint. Seldom do they cover solid technical matters in a clear or useful manner. Thus, they are of limited utility to the man who must go to the scene of the fire and come back with some answer as to what caused it, who set it, what defect was responsible, or generally, what happened.

It is the purpose of this volume to fill in some of these gaps—to explain the elementary technical considerations of the combustion process, of fuels, and of the investigative technique in simple enough terms so that the relatively untrained investigator or the attorney who must understand the implications of the investigation, will be aided significantly. We cannot, as yet, expect all investigators to be sufficiently conversant with chemical principles so as to be able to apply them effectively, nor can we expect all investigators to be expert in their operation. We can, however, hope to upgrade markedly the average performance as well as to increase the percentage of expert investigators. It is the author's hope that the information carried in this volume will assist in achieving these desirable goals.

PAUL L. KIRK

January 1969

1

Introduction

Fire losses are one of the most common causes of civil litigation and one of the most difficult areas in which to reach firm conclusions. The successful prosecution for arson is exceptionally difficult; similarly, civil cases also present facets of uncertainty which make unusual any definite and unambiguous outcome. Deliberately set fires, according to many investigators, are responsible for half of the number on which insurance claims are made. This figure is probably high; nevertheless, many fires are kindled for the purpose of collecting insurance. Most of the remainder, it is believed, are the result of carelessness, lack of adequate precautions, or other preventable causes. A significant number are the result of faulty equipment, especially electrical. However, this is a minor problem since these are usually in the preventable class.

From the standpoints of public safety and of the insurance carrier, it is important that investigative methods for determining the causes of fires be understood and applied to the fullest extent possible. Although fire has, in various forms, been known to man since prehistoric times, it is still poorly understood by many people, including some fire investigators, fire fighters, and others whose occupations expose them to the need for such understanding.

There is a natural tendency on the part of nearly everyone to be impressed by the magnitude and violence of the fire itself and the degree of destruction that follows a large fire. This tendency is aggravated in the case of insurance investigators by the fact that they will have to assess this damage and possibly pay a claim on it. For the investigator of fires, this preoccupation with magnitude and results of the fire is unfortunate. Cause not extent, should be his concern, and too much attention to the latter will obscure his understanding of the former. Because of these factors, and the frequent complexities of fire investigation, there is an urgent need to expand the approaches of the investigator.

To be a successful fire investigator, numerous facets of fires, fuels, people, and investigation procedures must be mastered and understood. The investigator must truly comprehend *how* a fire burns and that not all fires necessarily burn in the same way. These differences must be correlated with their causes, generally the nature of the fuels involved or the physical circumstances and environment of the fire.

One of the most important, and yet often neglected, areas of fire investigation and understanding is that of the fundamental properties of fuels, which will determine to a large extent the nature of the event that follows their ignition. Since these basic properties are determined primarily by chemical differences and the chemistry of the combustion itself, it is highly desirable that the fire investigator have a reasonably thorough knowledge of the relevant aspects of chemistry. The attorney is also not immune to this need, for if he has no understanding of what happens in chemical reactions and of their implications, he will be seriously handicapped in questioning witnesses, both lay and technical. He may even fail to recognize highly significant facts of the fire which can be instrumental in causing adverse judgments in court trials. This would also influence his decision as to when to try a case and when it should be settled on the best available terms. Thus, it is advantageous for both the investigator and the attorney to have some fundamental understanding of simple chemistry.

Since nearly all fires that enter litigation are either residence or industrial fires in structures, there is a tendency to think of fires in only these terms. This error becomes apparent when a forest fire is under consideration, or where the fire is of highly local, but very specialized, nature. Such fires are not uncommon when dealing with new products of modern technology with which even experienced investigators have no familiarity.

Explosions, closely akin to fires, and often accompanying them, also must be understood by the investigator and the attorney. The difference between the two types of events is often so small as to allow confusion and may be almost solely a matter of definition. The interrelation of the two types of chemical event must be clear for the proper interpretation of the combined phenomenon when it occurs. The explosion that accompanies the fire is not ordinarily very similar to that associated with bombs or industrial explosives. The distinction is sometimes lost both in the investigative and the legal phases of subsequent actions, although the investigation of a bombing is actually very different in character from that of the fire-associated explosion. These distinctions and comparisons will be developed in greater detail at various points in this volume.

Another curious tendency has been noted in connection with routine investigations of fires, namely, the tendency of some investigators, often associated with fire departments, to show an inherent bias as to cause. By this is meant that some investigators appear to believe that all fires have a single cause, for example, electrical equipment or smoking, and a vigorous effort is expended by these persons to prove that the new fire also had this cause, rather than to spend the time in seeking the true cause. Prejudgment of fire causation is as dangerous as prejudgment of any other unexplained event or crime. It is unfortunate that it appears at times to be an occupational disease of the investigative field, not only in connection with fires, but perhaps as noticeable there as in other aspects of investigation.

The complexities often encountered in fire investigation are sometimes overwhelming to the investigator, but patience and adherence to principles of combustion will generally allow a reasonable diagnosis of a fire. It is these elementary principles of combustion and of fuels, as well as the equally important principles of investigation as they apply to fires, that are the concern of this discussion. It is hoped that they will be considered valuable to investigator and attorney alike.

2

Elementary Chemistry of Combustion

Fire is a chemical phenomenon, accompanied by physical effects. The basic nature of each needs to be understood by the fire investigator, even though he may not have received a formal training in chemistry or physics. Since fire consists of a number of chemical reactions, occurring simultaneously, it is important to understand what a chemical reaction is and how it is involved in the fire.

All matter is made up of chemical compounds, some simple in composition, and others very complex. Most fuel materials fall in the latter class, although the net reactions of the fire do not involve the complexity of the material being burned to a serious degree. A *chemical compound*, represented by a molecule, is a combination of atoms of certain chemical elements in very definite proportions and arrangements determined by valence of the elements, combining capacities and similar considerations. The atom is the fundamental particle of an element. It is the smallest unit of matter that is involved in chemical reactions. Thus, the simplest atom, that of hydrogen, designated as H, cannot be broken into any smaller particle except by physical methods. Hydrogen atoms, in common with nearly all other atoms, will not long remain uncombined. If other elements are present there will be a tendency to combine with them to form compounds. If the hydrogen is pure, the atoms will satisfy this combining tendency by forming what is known as a diatomic gas, H_2. This signifies the combination of two hydrogen atoms with each other and represents as well the actual form of hydrogen gas.

Another gas, oxygen or O_2, is also a diatomic gas which makes up approximately 20 percent of the air. This gas is the most critical component of all practical fires, because it is essential for common combustion. Eliminating it will result in extinguishing the fire. Fire is not only a number of simultaneous chemical reactions, but overall, it is a series of oxidative reactions. This means that atoms contained in the fuel are combining with oxygen of the air, or becoming oxidized. There are many types of chemical reaction, but the major ones which occur in fire are oxidations. It is of interest to see what occurs when the excellent fuel, hydrogen, combines with oxygen or is oxidized. Two diatomic molecules of hydrogen combine with one diatomic molecule of oxygen to form two molecules of water. The chemist expresses this simple reaction by the equation

$$2H_2 + O_2 = 2H_2O.$$

Because water is a very stable compound as compared with the gases which form it, the reaction occurs with great vigor and the output of much heat. It is thus termed an *exothermic* (heat producing) reaction. If the gases are mixed before being ignited, a very violent explosion will result. When combined in a flame, an extremely hot flame is produced. Because hydrogen is an almost invarible component of fuels, and the hydrogen contained in the fuel in the form of atoms within complex molecules is converted to water during the oxidation, large quantities of water vapor are an almost invariable accompaniment of fires. This vapor will not be noticed unless it finds its way to cool surfaces, where it will condense and actually flow down such a surface in streams. The effect is often noted on tiles, mirrors, etc., in rooms of a building in which a fire is burning at enough distance that it does not heat up the condensing surface. The conversion of chemically bound hydrogen in the fuel into water vapor during the fire also invariably leads to the production of much heat, although less than when pure hydrogen is burned. This is because some of the heat energy is expended in breaking the chemical bonds that hold the hydrogen in the complex fuel molecule.

Another element, carbon, represented chemically as C, is of very special interest in connection with fires, because it likewise is an almost invariable major component of fuels, and in fact the element around which are built most of the molecules known as *flammable*. Although the diamond is composed of pure carbon, it exists in a special crystalline form which makes it difficult to burn. When it does burn, it gives rise to the same products, energy production, etc., as any other form of carbon. Better known, is graphite, the other common elementary form of carbon. This, also, because of its crystal form, is difficult to burn. When burned, it is consumed so slowly, that high temperature crucibles are often made from it. Charcoal and coke are both industrial products consisting of quite impure carbon but, nevertheless, primarily of carbon. These materials do not ignite easily. However, when afire, they burn with production of great heat and are consumed at a slow rate. The general chemical equation ordinarily applied to the oxidation of carbon is

$$C(solid) + O_2 = CO_2.$$

Carbon dioxide, or CO_2 is always produced in fires of carbonaceous material. It is an end product of nearly all combustions, including that which occurs in the animal body. The exhaled breath is markedly enriched with carbon dioxide formed by oxidation of foodstuffs in the tissues, just as occurs in the fire, although the foodstuffs are not carbon, pure or impure, but compounds of carbon with other elements. As a practical matter, in all fires there is also another reaction, which may be quite secondary, or may become primary depending on the supply of oxygen. It is

$$2C(solid) + O_2 = 2CO.$$

Carbon monoxide, CO, the product of this reaction, is the commonly known gas that produces an asphyxiating effect which will be dealt with in some detail later in this volume. While it invariably accompanies combustion reactions, it will not reach dangerous proportions in a properly adjusted gas appliance. It will be relatively high in a structural fire.

The three reactions given, although merely indicating the basis of the final combustion products, and without any attempt to define the complex mechanisms by which the products are actually formed, constitute the three most basic reactions of the fire. Water and carbon dioxide are the major products of nearly all fires with carbon monoxide at somewhat lower concentrations in the effluent gases. Some fuels, notably those of coal or petroleum origin, contain almost entirely carbon and hydrogen and give only small amounts of other gases. A possible exception is sulfur which is an impurity in most raw fuels. It also is oxidized to form sulfur dioxide, according to the following simple equation

$$S(solid) + O_2 = SO_2.$$

Sulfur dioxide, SO_2 is the very sharp smelling gas often noted around metal smelters and other industrial installations where sulfur is frequently oxidized.

Numerous other elements are found in the types of fuel most often the source of concern to fire investigators, e.g., nitrogen. This element does not burn in the sense of generating an exothermic reaction. Its fate in the fire is both complex and inconsequential as a fuel material. The same can be said for numerous inorganic elements present in many fuels, such as wood. These are the constituents that form the ash that remains. They do not contribute significantly to the combustion itself, with certain exceptions that will be mentioned later in this volume.

FUEL COMPOUNDS

The compounds that form good fuels are legion, but they fall into relatively few classes. Preeminent as a class are the *hydrocarbons*, compounds composed solely of carbon and hydrogen. The simplest of these is *methane*, CH_4, which is the chief component of natural gas. Written structurally,

$$\begin{array}{c} H \\ | \\ H - C - H \\ | \\ H \end{array}$$

indicates the important fact that carbon has four valences or combining points on each atom. Hydrogen has only one, which means that four hydrogens will combine with a single atom of carbon. In addition, carbon is somewhat unique in that it has a very strong capacity for carbon atoms to combine with each other to form chains, rings, and other complex structures. This might be illustrated by the compound, butane, which is packaged as a liquid petroleum gas, whose formula is C_4H_{10}, and written structurally is

$$\begin{array}{c} H \quad H \quad H \quad H \\ | \quad | \quad | \quad | \\ H - C - C - C - C - H \\ | \quad | \quad | \quad | \\ H \quad H \quad H \quad H \end{array}$$

Such chains can be branched to lead to more complex structures without including elements other than carbon and hydrogen. Since there are many possibilities for the location of such branching, a large variety of compounds may exist which have the same empirical formula but different structures. This can be illustrated by a compound such as isobutane, with the same empirical formula as butane but the structure,

$$\begin{array}{c}\text{H} \quad \text{H} \quad \text{H} \\ | \quad\;\; | \quad\;\; | \\ \text{H}-\text{C}-\text{C}-\text{C}-\text{H} \\ | \quad\;\; | \quad\;\; | \\ \text{H} \quad\; | \quad \text{H} \\ \text{H}-\text{C}-\text{H} \\ | \\ \text{H}\end{array}$$

Both chain length and degree and point of branching can be extended almost indefinitely, so that there is an enormous number of possible compounds, all of which are relatively simple hydrocarbons.

In addition to this type of compound, known as aliphatic because there are no rings in their structure, there are the aromatic hydrocarbons. The simplest compound of these is benzene, C_6H_6, with the structure

This compound also can have branches, any one of which may be very large and complex. Again, the simplest of the branched type of compound is toluene, C_7H_8, with the structural formula

It will be noted that the benzene ring in the two compounds above have been written with double bonds between some of the carbon atoms. This represents the fact that not all of the available four atoms of carbon are satisfied by single combinations to other atoms. While these may be considered as resonating bonds, rather than the fixed ones indicated in the formula, this unsaturated condition is typical of all the benzene ring compounds, of which there are a multitude. From the standpoint of combustion, this represents an excess of carbon over hydrogen in these "aromatic" compounds as compared with the type of aliphatic compound illustrated. This leads to a more orange flame, generation of more soot than a compound containing the same number of carbons in the saturated aliphatic hydrocarbons, and a slightly lower heat output. This is partially because hydrogen oxidation generates more heat proportionately than does carbon upon combustive oxidation.

Because the aliphatic compounds listed were all "saturated" or lacking double bonds does not mean that all aliphatic compounds are in this category. For example, the gas ethene (not ethane which is saturated) is C_2H_4, written structurally as

$$\begin{array}{c} H \\ \diagdown \\ C = C \\ \diagup \diagdown \\ H H \end{array} \begin{array}{c} H \\ \diagup \\ \\ \end{array}$$

while propene, C_3H_6, is

$$H - \underset{H}{\overset{H}{\underset{|}{C}}} - \underset{H}{\overset{H}{\underset{|}{C}}} = C \overset{H}{\underset{H}{\diagdown}}$$

All hydrocarbons are good fuels, but not all are commonly found as pure compounds. In fact, it is relatively expensive to isolate from petroleum or coal tar mixtures any compound in a pure state. Thus, virtually all such materials associated with fires are mixtures of relatively large numbers of individual compounds, but with sufficient similarity of chemical structure that their behavior in the fire may be quite similar. Thus, gasoline, one of the most common and best known, routinely contains close to thirty compounds, all hydrocarbons, in a liquid mixture. Most of these are aliphatic, or straight chain compounds, and most are saturated (no double bonds). To generalize further requires much more extensive discussion which is of no serious import to the fire investigator. For practical purposes, the mean properties of gasoline, insofar as calculations concerning combustion are concerned, may be taken as being similar to octane, C_8H_{18}, although little straight chain or normal octane may actually be present.

Perhaps the most important type of organic compounds concerned in the study of combustion processes is in a different group, known as the *carbohydrates*. These make up the bulk of wood, which is the commonest fuel of structural fires. They differ from the hydrocarbons in very significant ways.

Here, the molecules are very large and complex. More significantly, they contain a relatively high content of oxygen; that is, *they are already partially oxidized*. The process of burning wood is simply a completion of the oxidation that started in the synthesis of the fuel itself.

Carbohydrates are so named because their chemical formulae include the elements carbon, hydrogen, and oxygen in multiples of the simple formula

$$CH_2O$$

which is a combination of a carbon with water (hydrate). Actually, the simplest carbohydrate has the formula

$$C_6H_{12}O_6$$

which is the empirical formula for several so-called monosaccharides, of which the most common is glucose, the sugar found in blood and obtained from grapes and other fruit. It is sometimes called grape sugar. Cellulose, the major constituent of wood, and therefore the chief material that is consumed by fire, is made up of many units of glucose attached to each other in chains. Since the number of glucose units in the cellulose molecule is indeterminate, the formula must be written

$$(C_6H_{12}O_6)_x$$

where x stands for some rather large, indeterminate number. Since cellulose is simply a replication of glucose units, the main reaction that occurs on burning of the cellulose is the same as that for glucose

$$C_6H_{12}O_6 + 6O_2 = 6CO_2 + 6H_2O.$$

As with other fuels, not all of the carbon is ordinarily oxidized to carbon dioxide. Somewhat less than the indicated amount of oxygen may be used and a portion of the CO_2 replaced by CO, carbon monoxide. It will be noted that there is no effective oxidation of hydrogen, because in the balance, it has already been fully oxidized. This partially accounts for the fact that wood fires do not readily achieve the high temperatures of many other fuels. In this same connection, it should also be noted that other constituents of the wood, such as resins, may generate considerably more heat than will equivalent amounts of cellulose. Resinous woods are likely to produce hotter fires than non-resinous.

The primary effect of the heat on wood is to pyrolyze it, a process which produces much water and leaves behind charred wood which is primarily carbon or charcoal. This burns with considerably greater heat output than does the original wood, since it is with the carbon that the primary oxidation occurs in this instance.

It is apparent from the above considerations that fuels differ among themselves, not only in their chemical nature but in their mode of oxidation, the heat output of the reaction, and in other manners. It is also true that they have different characteristics with respect to the temperatures at which they can ignite, the quantity of oxygen utilized per unit amount of fuel, and several other related types of property which will be discussed under the subject of properties of fuels.

The simple equations that have been given for oxidation of several components of fuel system may actually carry much more information than was indicated. Gaseous materials, whether originally gases or derived by heating of liquids or solids, burn exclusively in the flames.

Thus, we may write the chemical equations for such fires in some such form as follows:

$$CH_4 \text{(methane of natural gas)} + 2O_2 = CO_2 + 2H_2O.$$

An equation such as this is a shorthand way of saying that one molecule of the methane, which is the major component of natural gas and also derived from solids by pyrolysis, combines with two molecules of oxygen to yield one molecule of carbon dioxide and two molecules of water vapor. It also tells us that one volume of natural gas will combine with two volumes of oxygen to produce three volumes of the products, if measured at the same temperature and pressure. Since approximately one-fifth of the air volume is oxygen, it also means that ten volumes of air are required to burn one volume of natural gas. This relationship is of the greatest importance in considering the facts and conditions of a natural gas fire.

Consider now the combustion of octane, a constituent of gasoline and some other petroleum fuels. The corresponding equation is

$$C_8H_{18} + 12.5O_2 = 8CO_2 + 9H_2O.$$

This equation tells us that one molecule of octane vapor requires 12.5 molecules of oxygen to oxidize it fully with production of 8 molecules of carbon dioxide and 9 molecules of water vapor. It also tells us that one volume of octane vapor requires 12.5 volumes of oxygen or 62.5 volumes of air to produce 17 volumes of gaseous products, i.e., 1.6 percent by volume of the vapor in air is the ideal combustion ratio. Octane therefore is a much better fuel than methane, since one volume of it leads to about six and one-half times as much combustion as one volume of natural gas, with correspondingly much greater heat production, air consumption, and product volume. It should, of course, be remembered that the equation represents complete combustion, which may not be obtained in an actual fire, and that all volume relations are reduced to the same temperatures.

STATE OF THE FUEL

Gases, liquids, and solids have been mentioned. The state in which the fuel exists must be related to its other combustion properties. There are relatively few gaseous fuels at ordinary temperatures. Hydrogen and methane have been discussed. However, many materials become gaseous at temperatures such as those existing in a fire, and virtually all can be converted to the gaseous state by applying sufficiently high temperature to it. As a very practical illustration, gasoline was discussed briefly above. Gasoline, and in fact all liquids, will fail to burn when in the liquid state, but liquids are generally easy to vaporize, and the gas that is formed burns like any gas, by mixing with oxygen and combusting as a flame.

In the group of hydrocarbons there are a number of materials that are difficult to classify as gases or liquids. The liquid petroleum gases, for example, of which propane and butane are the prime examples, can exist briefly in both states at ordinary temperature and pressure. Because they are so very volatile, when exposed, they readily vaporize, leaving no liquid. This is more true of the less complex molecular material. Propane, for example, vaporizes much more rapidly than butane, which is heavier, more complex, and has a higher boiling point. When gases such as these are placed under pressure, they condense to form liquids; the propane and butane of commerce, available in tanks, are largely in the liquid form. On releasing the pressure, as in feeding the overlying gas to a burner, more of the liquid vaporizes, and maintains the same pressure (critical pressure) in the tank until all of the gas has evaporated.

Gasoline, being still more complex in its structure and having heavier molecules, has a higher boiling point and is mostly liquid at ordinary temperatures and pressures. Thus, it is not necessary to seal it into a tank to retain it as liquid. When an opening or vent to the exterior of the tank is present, it is small and allows only minor escape of the gaseous portions of the gasoline. In order to combust in the cylinders of an automobile, most of the gasoline is actually vaporized in the carburetor and the remainder is vaporized in the hot cylinder, which means it is a totally gaseous reaction of combustion. With a cold motor, in which some of the gasoline may not vaporize, the portion that remains liquid is not combusted and tends to escape past the pistons into the crankcase. Such considerations may appear very elementary, but if overlooked in the interpretation of a fire, they can lead to incorrect conclusions.

Solids, as will be discussed in greater detail in another section, may burn, although as a rule they do not. When a solid is burning, it also is glowing on the surface. Yet the non-glowing solid may be surrounded by flames, and we say that it is burning. This is not strictly true in the absence of a glowing surface. Instead of the solid, it is strictly the gaseous pyrolysis products, resulting from heat decomposition of the solid, that are burning. The glowing phase will generally follow only after pyrolytic decomposition of the solid has effectively ceased, and a charred surface remains which is hot enough to continue a surface combustion in which two phases, solid (char) and gaseous (air) interact.

Perhaps the most important concern of this short discussion of state of the fuel relates to the fact that the classification of liquids, solids, and gases is far from absolute. They are to a large extent interchangeable, and the conversion is a function of the temperature and pressure existing.

Supplemental References

Burtsell, A. T. "Chemistry for the Fire Fighter." *Fire Engineering*, 119, 54, 1966.
Shepperd, F. "Physics and Chemistry of Fire." In: *Fire Chief's Handbook* (2nd. Ed.), D. M. O'Brien, Ed., Chap. 9, Reuben H. Donelley, New York, 1960.
Shorter, G. W. "Chemistry of Fire." *News Letter, International Assoc. of Arson Investigators*, 5, 48, 1955.

3

The Nature and Behavior of Fire

As indicated in the last chapter, two types of combustion (or fire) are: (1) flaming and (2) glowing. The first is a gaseous combustion in which both the fuel and the oxidant are gases. The second involves the reaction of the surface of solid fuel with gaseous oxidant (ordinarily oxygen of the air). Nearly all fires are primarily of the first type. However, the second is not uncommon, either alone or in combination with the first. For example, the "smoldering" fire in a mattress or in a pile of sawdust is a good illustration of the glowing fire, as is also the charcoal fire used for barbecues.

The differences are a result of the nature and condition of the fuel, as well as the availability of oxygen. If the mattress or sawdust is stirred up, it may develop into a flaming fire.

On the other hand, the glowing charcoal has little or no flame because the compounds that could be rendered volatile by heat were lost at the time the original fuel was charred or coked. The differences between a wood fire and a fire of charcoal made from the wood are discussed later in the section dealing with the properties of fuels.

In order for a fire to occur, several conditions must exist. These may be stated as follows:

1. Combustible fuel must be present.
2. An oxidant (e.g., oxygen of air) must be available in sufficient quantity.
3. There must be adequate ventilation.
4. Some means of ignition by application of heat must be applied.

Although these requirements appear obvious, it is true, nevertheless, that statements have been made in court describing fires burning on bare concrete floors devoid of fuel. Absolute lack of any means of ignition is often claimed. In general, disregard of the simple requirements stated is not at all uncommon. Furthermore, the conditions of some fires are so uncommon as to require special analysis of the factors given.

Experience with the common gaseous, liquid, and solid fuels is so widespread that few misjudgments would be made regarding their presence and availability for burning. However, there are many other less common materials (e.g., plastics and metals), involved in some fires, about which experience is less helpful. In addition, the mere presence of a suitable fuel in conjunction with the other conditions listed does not guarantee that a fire will result. For example, common furnace oil is excellent fuel, but when spread on a concrete slab, it will generally resist every effort to ignite it, even with a blowtorch. Questions such as these require detailed knowledge of the properties of the fuel and of the character of fire itself. Another little-known fact is that newspapers, which are so helpful in kindling fires, may readily be used to extinguish certain types of small fires. Thus, availability of suitable fuel and the judgment as to what fuel is suitable in specific instances are matters requiring more than ordinary experience to assess.

FLAMING FIRE

A fire characterized by flame is the most common type. Here, the flame actually *is* the fire. An important fact to remember is that there cannot be a flaming fire unless a gas is burning. This holds true whether the gas is preformed, evaporated from a liquid, distilled, or pyrolyzed from a solid. The flame is a totally gaseous reaction. With liquid fuels, only flaming fire can result, since liquids per se do not burn. The reactions of the fire are, under ordinary circumstances, oxidative, with oxygen of the air as the oxidant.

Solid fuels often give flaming fires predominantly, although they are frequently accompanied by glowing fire. Thus, wood tends to burn with flames, especially in the early stages of the fire. Later, the charcoal formed in the initial fire will continue to combust as a glowing fire. Coal, on the other hand, tends to give a much lesser amount of flame, and proportionately more glowing combustion. This corresponds with the fact that heated coal gives off limited quantities of volatile gases, as compared with wood.

GLOWING FIRE

A glowing fire is characterized by absence of flame but presence of red hot materials on the surface of which combustion is proceeding. The charcoal fire is a good illustration, while coal also often approximates a glowing fire. In the past, it was common to see the blacksmith heating iron objects in a fire made with coal through which a forced air draft was passing. These fires showed very little flame, but the glowing fire induced by the air draft was intensely hot—more so than most flaming fires.

Since the glowing fire involves two phases, a solid and a gas, the ratios of oxidant and fuel are not of direct concern, although the chemistry of the reaction still determines the total consumption of each. In this instance, unlike the flaming fire, they are not premixed, or even mixed at the moment of combustion, but react at solid surfaces over which the oxidant gas passes. Inasmuch as fuels are generally organic materials, which are often reduced to charcoal (or coke) as the volatile decomposition products are distilled out and burned, most fires reach a glowing state in their later phases. It must be remembered that the glowing material here is not the original fuel, but a derived fuel, charcoal or coke, resulting from the earlier portion of the fire.

An important feature of the glowing fire, especially when there is a forced draft as in the blacksmith's forge, is that the limited flame that accompanies it is chiefly from carbon monoxide burning to carbon dioxide. The carbon monoxide is generated by the oxygen of the air reacting deficiently with the excess of hot carbon, from which this gas rises and is further oxidized with a hot but rather small, generally bluish flame.

In the investigation of most fires, the glowing phase is not of great consequence because it generally represents the last part of the chemical process. The investigator, and indeed the fireman, is concerned much more with the earlier aspects when flames are involved. The fireman is interested in extinguishing all glowing fire, because from it may come reignition of remaining combustible material. The investigator is primarily concerned only with the initial small fire that grew into the large one.

Some exceptions to these rules must be considered. A fair proportion of active fires follow a period of smoldering, this being essentially a glowing fire. When the fire is kindled in an environment in which ventilation is very limited, it does not generate the heat necessary to burst into flames. This slow combustion may proceed for an appreciable time before the fire erupts or is discovered. Such fires are found as a result of ignition of mattresses or upholstery padding in residences, haymows in barns, and sawdust piles in mills. To them are attributed numerous fires that apparently start at times when there is no known source or likelihood of ignition. It is assumed that they started earlier, "smoldered" for a period of time, and finally burst into flame. It may be noted that arson attempts sometimes fall into this category. This is because the fire is kindled in a restricted space where the available oxygen supply is deficient, and flames are suppressed, but enough air is available to allow a smoldering fire to continue.

Another consequence of smoldering is from the discarding of previously flaming but supposedly extinguished materials which are still glowing. Discarded matches which are not specially treated will continue to glow for some time after the flame is extinguished, and can initiate a fire. Certainly the glowing cigarette falls into this category (see later section). Hot coals, spilled from fireplaces, barbecues, and other similar sources will start larger fires if they contact suitable fuel.

EXPLOSIVE COMBUSTION

The explosive combustion is not normally classified as a fire. It is sometimes termed an "explosion" although it is not the same as true explosive behavior. It requires consideration because this type of combustion does accompany fire, often as an initiating factor and sometimes during a conventional fire when favorable circumstances develop. Since this type of combustion is considered in a later chapter, it will be discussed only briefly at this point.

Explosive combustion occurs only when vapors, dusts, or gases, premixed with an appropriate amount of air, are ignited. Under these circumstances, the combustion that results is not different from that which happens in the burning of these materials, except that the premixture allows the entire combustion to occur in a very short space of time. Thus, all of the heat, development of combustion products and their expansion, which normally would require an appreciable time, becomes an almost instantaneous event and is recognized as an explosion. The event may be very forceful and produce great damage, including the blowing apart of an entire building, or it may be so small as to be only an audible pop. Crackling of flames, and the pops that occur in a wood fire in the fireplace are similar in nature, although not always due to the same cause.

At times, the first phase of a fire is an "explosion." What has occurred is that flammable gases or vapors have accumulated from some source, mixed with appropriate amounts of air, and become ignited. Such preliminary explosions are often followed by flaming fire. In such instances, it is the cause of the initial explosion that is the concern of the investigator, since he knows that flammable gases were released from some source and that they were later ignited.

When an explosion of this type occurs during a fire, it is an indication that a new source of fuel was made available locally and in some quantity. Perhaps the most common source of this material is a sealed can, bottle, or drum of a flammable liquid which has broken or burst because the heat of the fire expanded the contents beyond the strength of the container. Being hot, this material vaporizes very rapidly, and produces a local explosive environment just before it ignites. From the investigative standpoint, this event has two values: it proves that the container was sealed initially, and that it was not causal in producing the fire. It also proves that an extensive fire and much heat surrounded the container, and this situation is either the result of an unusual type of accident, or it is the work of an arsonist. These facts can be of utmost importance in the interpretation of a good many fires.

HEAT

Aside from the significance of the chemical reactions which produce it, the most fundamental and important property of the fire is *heat*. Heat initially starts a fire, and the fire produces heat. As will be discussed under *ignition*, every method by which a fire may be ignited involves the application of heat, and nothing but heat is ever necessary to start the fire when the environment is suitable. *Heat, basically, is the only important consideration in starting the fire.*

On the other hand, *all of the destructiveness of a fire is the direct result of the heat generated.* Heat produces the damage to the structure, intensifies the fire, is the means by which the fire spreads and enlarges, and provides the greatest barrier to the extinguishment of the fire. When the heat can be properly studied and understood in connection with a fire, the sequence and cause of that fire will generally be very clear. There are several special considerations of heat as applied to a fire investigation that require understanding:

1. Heat as it applies to igniting the fire. This is developed further in the chapter on ignition.
2. Heat as it applies to the increase of the rate of chemical reactions (including fires) as developed chemically and expressed as the Q_{10} *value*.
3. The transfer of heat as the factor controlling the spread of fire when the additional conditions of available fuel and oxygen are met.

Heat in connection with igniting fires is part of a broader consideration (presented as a separate chapter later in this volume) of the means of producing and applying the heat.

The Q_{10} Value

The Q_{10} *value* is derived from the chemical generality that the rate of all chemical reactions shows a dependence on temperature. With rare exceptions, all of them have a higher rate the greater the temperature. The Q_{10} value is the increase in rate of the reaction which results from raising the temperature 10°C. (18°F.). For most reactions, its value is two or more. Its great importance stems, at least in part, from the fact that the fire generates much heat and raises the temperature of the reacting components, thus increasing the rate of reaction. This, in turn, generates more heat, thus again increasing the rate of reaction. Were it not that the diminishing availability of fuel and oxygen, combined with loss of heat to the larger surroundings, to bring this chain reaction under control, every fire would become a violent and rapid holocaust. Since these factors do act to control the fire, the most immediate significance of the concept is in connection with those processes which lead to spontaneous combustion. This subject is treated separately in a later section. However, it can be briefly considered in the following light: Assume that a system of fuel and oxygen is properly contained, insulated and proportioned, and a very limited exothermic reaction is proceeding. This might be a slow oxidation, bacterial activity or some other type of chemical transformation in which heat is generated. If the heat cannot escape as fast as it is generated, the temperature will rise. As it rises, it speeds up the very reaction that is generating the heat, thus leading to an increasing rate, resulting in greater and greater quantities of heat. This heat will finally be sufficient to raise the temperature of the system to the ignition temperature of at least part of the fuel. Naturally, when this temperature is reached, the fuel ignites to a free burning fire, and we can say that it resulted from spontaneous combustion.

Heat Transfer

An integral part of every fire is the *transfer of heat*, both to the fuel which is critical to continuity of the fire, and from the regions of combustion, whatever their type. The principles of heat transfer are relatively simple and are vital to the understanding of the fire itself.

Heat is energy in a kinetic form, or energy of molecular motion. Except at absolute zero (−273°C.), all matter contains heat, because its molecules are in movement, both translational and vibrational. Temperature is merely an expression of the relative amount of this energy. The total amount contained in any given quantity of matter is conditioned also by the heat capacity, which in turn is related to such properties of the matter as its density, form (gaseous, liquid, solid) and other physical properties.

Heat is transferred in three ways: (1) *conduction*, (2) *convection*, and (3) *radiation*. All of these methods are operative in fires. However, the relative importance of each will vary with the intensity and size of the fire, as well as with the configuration of the environmental system that is burning. Each will be considered separately.

Conduction. Conduction is the transfer of heat energy through a medium by virtue of molecular activity. Since actual contact of the vibrating molecules is necessary, conduction is limited to a localized sphere of action. Its consequences are most noticeable in solid materials where convection does not occur. It is readily illustrated by heating one end of a rod, and observing the temperature behavior of the other end. Heat traveling through the rod by conduction will cause a rise in temperature at the other end, but with considerable lag. The amount of heat flowing through the rod is proportional to the time, the cross-sectional area, and the difference in temperature between the ends, and is inversely proportional to the length. It also varies with different materials. The relation may be expressed by the following equation where appropriate symbols are used for each factor:

$$H = k \frac{A(t_2 - t_1)}{L} T$$

where H is the amount of heat flowing through the body of length L and cross section A, k is the thermal conductivity, T is the time interval of flow, t_2 is the temperature of the hot end, and t_1 is the temperature of the cold end.

An illustration of the importance of thermal conduction might involve the relative behavior of a piece of metal, e.g., a wire, or a galvanized metal sheet as compared with the behavior of a piece of wood in the same fire. The thermal conductivity of metals is high so that if the metal piece is heated, the heat rapidly spreads to unheated areas. If the temperature in these portions rises above the ignition temperature of any fuel material in contact with the metal, this fuel will be ignited at a distance from the initial source of the heat. A piece of wood, on the other hand, being a poor conductor of heat may burn and be strongly heated for a long interval. The heat, however, does not spread through it well, and it will not be expected to ignite at a distance. One side of a simple board will show deep charring in a fire, and the opposite side, unless exposed to flame from another source, will remain uncharred and apparently normal wood.

The importance of k, the thermal conductivity which relates to the material through which the heat is being conducted, in the development and consequence of fire is considerable. Not only is it involved in the ignition of materials, in which heat must flow from the flame or spark to the fuel and to some degree through it before a fire will start, but it also has a marked effect on the degree of fire damage. This can be better illustrated by comparing some conductivities of actual materials often involved in fires (Table 1).

The value of k is defined as the quantity of heat in calories that will pass in 1 second through a 1 cubic centimeter cube when the two opposite faces are maintained at 1°C. difference in temperature. Regardless of the exact terms in which the numerical value is expressed, the relative rate of transfer of heat by the solid is shown by the relative figure. It will be noted that heat conductivity of some common fuels such as paper and wood is extremely low, while metals

TABLE 1. THERMAL CONDUCTIVITY OF VARIOUS MATERIALS

Substance	k
Copper	0.92
Aluminum	0.50
Brass	0.26
Iron	0.16
Glass	0.0025
Tile	0.002
Water	0.0014
Wood	0.0005
Paper	0.0003
Felt	0.00004

have relatively high conductivity. Copper, for example, conducts heat almost two thousand times as efficiently as does wood. This explains several of the commonly observed facts of fires. For example, a short circuit in electrical wiring may melt the insulation on the wires for a considerable distance from the short circuit, because heat is transferred so effectively through the copper wire. It also explains why one face of a section of timber may be burned very heavily, while other parts, not exposed directly to the fire, will be completely undamaged. Materials such as wood, and even more so, felt, are recognized as good heat insulators because of the property of transferring heat so inefficiently.

Convection. Convection, as a means of transferring heat, is extremely important in fires. It consists of mechanical movement of masses of liquids or gases and does not apply to solids. Thus, when water is heated, it can be rapidly raised in temperature despite its very low heat conductivity. This occurs because the heated water on the bottom of a container is expanded, becomes less dense, and is displaced by the heavier cold water which is heated in turn. Another very familiar example is in the heating of buildings by furnaces or heaters. Here, the fire is small and controlled, but the heated air coming from it is often circulated exclusively by convection. Only in forced draft furnaces is there any deviation from this principle.

In fires, the moving masses of hot materials are the gaseous products of combustion, along with surrounding air which is also heated. These become expanded and lighter and are displaced upward at a rapid rate. In large fires, especially outdoors, the upward movement of hot gases is often so great as to form what is sometimes called a firestorm. It is this form of heat transfer that accounts for most heat movement in a fire. This also largely determines the fundamental properties of fire with respect to its movement, spread, and ultimately its pattern.

Radiation. Transfer of heat by radiation is less commonly understood or appreciated than transfer by the methods discussed above. It is also less important in fire investigation than the other two methods of heat transfer. However, it is a factor of great consequence in some fires, especially the larger ones, and should be understood. Burns are sometimes not accounted for, when caused by radiant heat effects.

All objects radiate heat constantly, the radiation being exactly like light except that it is not visible to the eye. Radiant heat energy travels at the same rate as visible light, 186,000 miles per second. Radiation from the sun is the primary source of all heat on the earth. As soon as the sun is visible, the heat it radiates can be felt, just as the observer feels the radiated warmth from an open fire. Such warmth is felt because it is absorbed by the body, and the absorption is greater than the radiation of heat by the body under these circumstances. In the same way, a person in a cold environment is chilled because his body radiates heat faster than it receives radiated heat from the environment. In surroundings in which everything is at exactly the same temperature, all objects are both radiating and absorbing radiant heat, but at the same rate the result is no change in temperature. Thus, it is apparent that the rate of transfer of heat energy by radiation from one body to another is proportional to the difference in temperature of the bodies.

Temperature differences alone do not determine the rate of transfer of radiant energy. This rate is also subject to the nature of the exposed surfaces, and especially to the degree to which they reflect the heat rays. Moreover, objects that radiate heat readily also absorb it readily. Dark-colored objects both absorb and radiate heat better than light-colored ones. On the other hand, polished surfaces, such as chrome plate, resist both absorption and emission. The same mirror that reflects visible light also reflects the infrared, or heat rays, and in the same manner.

In regard to fire, several items of consequence may be mentioned. Radiant heat is a major factor that makes fire fighting difficult, because close proximity to the fire makes operations difficult. When a fire is burning in a portion of a structure, all surfaces that face the fire will be heated by the radiant heat, and because of their temperature increase, become susceptible to easy ignition. When this temperature reaches the ignition stage of the material itself, that material will burst into flame. In very large fires, it is not uncommon for adjacent buildings to be ignited at a distance solely by the radiant heat from the intense fire. This effect of radiation of heat is a large factor in the spreading of fire after the initial stages, with which the investigator is most concerned.

To illustrate the effect of radiation in spreading a fire locally, assume that an enclosed area such as a room has an intense fire burning in one side only. The hot gases have an escape close to the top of the enclosure, so they cannot carry the heat to the other side of the room and start the process there. Radiant heat from the local intense fire is, however, being transferred to combustible material on the other side of the room, and heating that material. As soon as it starts to produce flammable gases pyrolytically from being heated, these gases accumulate until some of them either reach the ignition temperature from the radiant heat itself, or physically move to an area in which flame is present. Thereafter, the entire mass of flammable gas undergoes a minor explosion, and the fuel from which it came is instantly on fire. This phenomenon is common and well known to firefighters. It is disconcerting to have objects at some distance from the active fire suddenly "explode" into flames, possibly trapping the men involved, and certainly adding to the total size and difficulty in fighting of the fire. Charred areas in places inconsistent with the ordinary fire spread are commonly due to this effect of radiation of heat, and should be so recognized by the investigator. Generally, they are not highly localized as is true often of direct spread of flaming fire. They are usually rather uniform in degree of charring and tend to violate the principle elaborated elsewhere in this volume that the low point of a fire is generally its origin. This is because radiant heat can also produce charring at low levels, even when the fire that generates the radiated heat is not at low level.

EFFECTS OF ENVIRONMENTAL CONDITIONS

The investigator of fires will be only mildly concerned with the role played by such environmental conditions as the temperature or humidity that preceded the fire, because the fire has already occurred. However, there are relationships to such factors that need to be taken into account at times. It is well known that when an area has become excessively hot and dry, there is a greatly increased hazard of forest and grass fires, and structural fires may be more common or of greater severity. Because of such a relationship, there is also a tendency to overemphasize it in many instances in connection with the fire that has already occurred. Wetness, an environmental condition which is subject to rapid alteration in a local region (e.g., wetness from fire hoses) is also a real factor in the ease of burning, the danger of kindling, and the intensity of the kindled fire. Although as stated, these are secondary to the investigation of a fire that has already occurred, they merit careful consideration by the investigator.

Temperature

Fires are more likely to become severe on a hot rather than a cold day. However, some of the most destructive fires have occurred when the water from fire hoses has frozen on the firemen's ladders. The reason for the greater fire hazard on the warm day is easily explainable.

1. Heat dehydrates, and a heated environment is dryer than an unheated one. Dry materials kindle more readily than less dry ones, although the heat

from the fire itself is quite sufficient to dry out moist materials ahead of the flames when there is no renewal of the wet condition.
2. As developed elsewhere, the rate of a chemical reaction (such as fire) is approximately doubled with every rise of $10°C$. In a full blown fire, this factor in the initial environment is negligible, because the heat from the fire itself is so much greater than an ambient temperature at the beginning. However, the warmer the environment, the faster the reaction in the extreme initial portion of the fire when the fire's own heat is still not so overwhelming. In other words, it is easier to kindle fire with hot than with cold fuel.
3. Heat (or temperature) is related to humidity, a factor that will be discussed in the next section.

The variations of heat in a normal living environment has little influence on most fires simply because the variations are not great enough, nor the influence direct enough, to make more than a quantitative difference. This is especially true in the matter of structural internal fires, and much less true with the forest or grass fire.

Humidity

Firemen tend to be greatly concerned with humidity for the reason that they note a greater incidence of fires on dry days than on wet days. Again, this refers far more to exterior fires than to interior ones. When it is hot and dry for protracted periods, it is well known that forest fires become an extreme menace; on cool, humid days, there is far less hazard. The tendency to apply the same quantitative reasoning to interior structural fires is often a misguided one, although not always. The question as to the actual importance of humidity is not a simple one to develop, because it involves several secondary factors.

Humidity refers to the amount of water vapor in the air. At saturation, this is determined by the air temperature. The greater the temperature, the greater the amount of water that can be held by it without condensing into fog, dew, or rain. Thus, it is not so much the absolute humidity that is relevant as the relative humidity, which is the percentage of the saturation value that is actually present. A relative humidity of fifty percent means only that the air contains one-half of the saturation value for that temperature. However, if the temperature is raised without any change in moisture content, the relative humidity drops to a lower percentage, without actually changing the total moisture at all. For practical purposes, it is always the relative humidity that is quoted and of importance, because this is the best measure of the moistness of the air at any given temperature.

It was stated above that heat (or temperature) is related to humidity. The relation is complex. One relation has been discussed above concerning relative humidity. Another relation stems from the fact that heating of liquid water causes vaporization, with a corresponding rise in the absolute, and sometimes the relative, humidity. Cooling of the environment tends to remove water from it. To illustrate the point, several types of climatic conditions should be mentioned.

Dry cold. In northern latitudes, especially in the wintertime, this condition is frequently seen. It means that the temperature and humidity are low. Sometimes, but not always, the relative humidity is not low, which leads to wet cold.

Wet cold. This type of climatic condition results from rather rapid chilling of a moisture laden air, generally accompanied by fog and dew. In this case, the only removal of moisture is caused by forced condensation by the cold air, and the relative humidity remains high. It is contrasted with the dry cold, in which very cold temperatures rise significantly after chilling the moisture content to a very low value, and the humidity has not yet risen significantly.

Dry heat. This situation is characteristic of desert climates during many of the weather cycles. The primary reason is that there is little water to evaporate into the air which is very warm. Thus, the heat is great but the humidity, both absolute and relative, is low.

Wet heat. Tropical climates often provide this situation, where there is a great deal of exposed water available to evaporate under the sun's rays as well as a large amount of exposure to the sun. The heat rises, and with it the humidity as the available water is evaporated into the air.

Most places on earth approximate to some degree all four of the basic types of meteorological conditions at times. This depends on the vagaries of the weather and the other conditions mentioned. For example, a thunderstorm on the desert can produce for a time a wet heat, and during the summer, even arctic regions in which water becomes scarce can have a dry heat because of long days of sunshine.

The *effects of humidity on fire* stem from several basic factors. When objects are exposed for sufficient time to low relative humidity, they dry out and are, therefore, more combustible than when less dry. However, the difference in water content is not so great as generally believed; many materials, such as paper, vary within surprisingly small limits with ordinary changes in humidity. Perhaps the greatest effect from this standpoint is on living vegetation, a major fuel of forest fires. Plants not only dry out as would a sheet of paper, but much more important they "transpire" moisture from pores in the leaf structure. Despite the closing of stomata in dry weather, a leaf may lose more moisture proportionately than a sheet of paper when exposed to drying conditions. The difference is clear when it is noted that dry periods correlate with increased forest and grass fires, far more than with increase in primary structural fires.

Another effect that is rarely given its proper significance is the relation to small sparks such as result from static discharge and from impacts of materials such as a steel object with a rock. Static tends to accumulate excessively in dry periods. However, this dissipates by leakage into the atmosphere when the relative humidity is raised. The other type of spark is merely a little more vigorous in dry atmosphere because of less leakage of charge. It can more readily ignite dry than moist materials that it may contact.

Most important to understand, is the fact that once the fire is actually burning, the humidity by itself has little or no effect on it. The effects are virtually confined to the primary ignition, not to the magnitude of the ensuing blaze. The reason is rather obvious, namely that the heat generated by the fire is so great, that all humidity considerations of the fire environment are totally overshadowed. It is also true that when fuel in a humid environment is exposed for a considerable period of time to local heat, it is dried by that heat, regardless of the ambient humidity.

A still further factor in the matter of humidity relates to the fact that on many days when the humidity is low, the temperature is high. In such an instance the effects mutually support each other, and neither may dominate.

Wetness of Fuel

This factor is not the same as humidity, which relates to wetness of air. It is of far greater importance than humidity when the degree of wetness exceeds that which is in equilibrium with the air (controlled partially by humidity). An object that is exposed to air reaches equilibrium with it and is stated to be "dry" as long as excess moisture in the form of rain, dew, or water from an artificial source is not applied to it. While all "dry" fuel materials burn readily, wet fuel must first have the excess moisture evaporated before it can be raised to its ignition temperature. Since water boils at $212°$ F., and all ignition temperatures are much higher than this value, it is apparent why so much extra heat must be applied to the wet fuel to cause its ignition. It can be stated that as long as free water (not chemically bound in the fuel) is present on the surface, fire is normally impossible. This gives the basis for fighting fire with water. If water can be added faster than the fire can boil it away, the fire will be extinghished because the temperature is lowered. If the water boils away as fast as it is added, the fire continues. Some exceptions exist. Some fuels, such as magnesium, actually react with water at high temperature. Therefore, magnesium cannot be fought with water, because water on hot magnesium generates the excellent fuel, hydrogen gas, and increases the size of the fire spectacularly. Another apparent exception is the oil fire which is not extinguished by water. This is true because the water is heavier than the oil and sinks below it, while the oil surface continues to burn. This effect does not invalidate the rule of wet fuel but represents another added factor which negates the effect. If the oil could be covered by the water, the fire would extinguish.

Wetness of fuel must not be confused with the effect of humidity, because they are totally different even though related in some aspects, as in the formation of dew. Fog has a similar effect on burning, although not as great as dew. It also tends to moisten and accumulate on surfaces and to make them hard to ignite. However, the effect is less because the amount of coalesced moisture is also less. Fuel that is wet internally from absorption of moisture is even more difficult to burn than fuel moistened on the surface only. The principle is identical, however, in that the water boils out at a temperature far below the ignition point and cools the fuel until there is no more free water to boil. Because of its excess water content, there is little relation between a waterlogged piece of wood and a piece of wood that has stood in humid air until it is in equilibrium with it, i.e., "dry."

Supplemental References

Brown, F. L. "Theories of the Combustion of Wood and Its Controls." *Rept., U.S. Forest Prod. Lab., No. 2136*, Madison, Wis., 1958.

Fristron, R. M., "The Mechanism of Combustion in Flames." *Chem. and Eng. News*, p. 150, Oct. 14, 1963.

Johnson, A. and Auth, G., Ed. *Fuels and Combustion Handbook*, (1st. Ed.). McGraw Hill, New York, 1951.

4

Combustion Properties of Non-Solid Fuels

Clear understanding of fires must include a thorough familiarity with the properties of fuels as they relate to combustion. Although this statement is absolutely fundamental, there is no subject about which experienced fire investigators are so liable to err. This unfortunate situation has developed because most fire investigators are not chemists, and considerable chemical knowledge is essential to the proper interpretation of fires, which are strictly chemical reactions.

A related problem that troubles some investigators is the understanding of terms that apply to standard tests devised in the laboratory to define the properties of the fuels that are of industrial significance. It is the purpose of this discussion to clarify some of the concepts that are basic to the understanding of how materials burn, the conditions and limitations that apply to combustion of fuels, and the conventional methods of expressing the properties in terms of laboratory tests.

TYPES OF FUEL

Combustion properties have different significance, or perhaps no significance at times, when applied to various types of fuel. Thus, the concept of flash point, discussed later, is essentially meaningless when applied to solid fuels, although it may apply to volatile materials derived by heating the solid fuel. The relationship between the type of fuel and the type of property is clarified below.

Gases. Flammable materials found in the gaseous form at all reasonable ambient temperatures will burn whenever mixed with the proper amount of air. Flash point has no significance with such materials, nor has boiling point. However, the flammable range of admixture with air and the vapor density are important properties.

Liquids and their vapors. Vapors from liquids are the materials that directly support the flame over a liquid. To the liquid-vapor combination, flash point is applicable and important, as is the boiling point of the liquid. Ignition temperature, range of composition of air mixtures that are flammable, and vapor density are critical properties of the vapors formed from liquids.

Liquids, as such, so rarely burn that their combustion properties are of little or no interest except, as indicated, for their relation to vapors formed from them.

Solids. Most solids of interest burn only at the surface, as a glowing fire. If the vapors which can be distilled and pyrolyzed from some solids are disregarded, most of the properties mentioned above are not applicable. Thus, they have no flash point in the strict sense, no vapor density, no range of admixture with air, etc. They do have other physical properties of importance, often difficult to designate as related to combustion properties. However, they have very important chemical variations that are of direct consequence. The density of the solid may be of significance, as is its porosity and its melting point, if any.

The chemical properties of solids relate to the relative combustibility, volatility, and other similar properties of their constituents. For example, some woods have a high pitch content which makes them burn with greater intensity than the relatively pitchless woods. Some other woods have chemical (and physical) properties which make them extraordinarily resistant to fire. These matters are discussed later in greater detail.

FLASH POINTS

This term is basic to the description of many conventional fuels and is often of concern in the operation of internal combustion motors and flammable liquids, not used in this way, but a fire hazard nevertheless. A liquid fuel must be able to generate a vapor in sufficient quantity to reach the lower limit of flammability before it can burn. This does not mean that at this temperature it will ignite spontaneously, but only that it can be ignited by some additional source of heat. For example, the flash point of gasoline is given as about $-50°$ F. All that such a figure indicates is that a gasoline-operated motor will be able to function at a temperature down to $50°$ below zero—not that the gasoline will ignite spontaneously at such an extremely low temperature.

Flash point is determined by placing the fuel in either a closed or open cup and heating or cooling it to the lowest temperature at which a spark will cause a little flash to occur over the surface of the liquid. No fire results from such a test except the little flash of flame which immediately extinguishes itself. In polar regions, the flash point of the gasoline has a very significant relation to the operation of the automobile or tractor. It has very little relation to the danger associated with fire caused by the material. Some typical flash points are given in Table 1.

TABLE 1. FLASH POINTS OF REPRESENTATIVE SUBSTANCE*

Substance	Closed Cup	Open Cup
	°F	°F
Acetone	0	15
Amyl Acetate-n	76	80
Amyl Alcohol-n	91	120
Aniline	168	—
Benzene	12	—
Benzine (petroleum ether)	—	< 0
Butane-n	−76	Gas
Butane-iso	−117	—
Butyl alcohol-n	84	110
Butyl cellosolve	141	165
Camphor	150	200
Carbon disulfide	−22	—
Castor oil	445	545
Cellosolve	104	120

*Data taken from "Properties of Flammable Liquids, Gases and Solids," *Loss Prevention Bulletin* No. 36.10, Factory Mutual Engineering Division, Boston, 1950.

TABLE 1. FLASH POINTS OF REPRESENTATIVE SUBSTANCE *(Continued)*

Substance	Closed Cup °F	Open Cup °F
Denatured alcohol (95%)	60	—
Dichloroethylene	43	—
Diethylcellosolve	—	95
Diethylether	−20	—
Ethyl acetate	24	30
Ethyl alcohol	55	—
Ethylene dichloride	56	65
Eythlene glycol	232	240
Fuel oil #1	110–165	—
Fuel oil #2	110–190	—
Fuel oil #3	125–200	—
Fuel oil #4	150+	—
Fuel oil #5	150+	—
Fuel oil #6	150+	—
Gasoline	−50	—
Glycerine	320	350
Hexane-n	−7	—
Kerosene (fuel oil #1)	110–165	—
Linseed oil	435	535
Lubricating oil (mineral)	—	300–450
Methyl alcohol	54	60
Methyl cellosolve	107	115
Methyl ethyl ketone	30	—
Mineral spirits	85	110
Naphtha (coal tar)	100–110	—
Naphtha (Stoddard solvent)	100–110	—
Naphtha (V.M. and P.)	20	—
Octane-n	56	—
Octanes-iso (commercial)	10	—
Paint liquid	0–80	—
Petroleum, crude	20–90	—
Petroleum ether	−50	—
Pine pitch	285	—
Propyl alcohol-iso	53	60
Pyridine	68	—
Resin oil	266	—
Rubber cement	50 or less	—

(Continued)

TABLE 1. FLASH POINTS OF REPRESENTATIVE
SUBSTANCE (Continued)

Substance	Closed Cup	Open Cup
	°F	°F
Rum	77	–
Soya bean oil	540	–
Toluene	40	45
Tung oil	552	–
Turpentine	95	–
Varnish	<80	–
Whiskey	82	–

IGNITION TEMPERATURE

The ignition temperature is basic to all considerations of the initial source of a fire. This is the temperature at which a fuel will spontaneously ignite. For example, gasoline with a flash point of $-50°$ F. requires a temperature of about $495°$ F. to catch on fire. This is the minimum temperature that must be reached by the match, spark, lighter, or other igniting instrument in order for any material to burn. With one or two exceptions, such as catalytic action, the temperature must be raised to this value, at least locally, before fire can result from any fuel. The ignition temperatures are generally so high as to rule out spontaneous combustion, except for a very small category of materials and only under very special conditions. Catalytic oxidations are exceptions but are very rare, generally only possible under contrived conditions. For example, platinum in a finely divided state may serve to allow combustion of certain flammable gases in air at low initial temperatures. In a fuel cell this may be important. In a fire, however, it is so rare as to require no special consideration.

Virtually all fires originate because there is some local high temperature in a region in which an appropriate fuel-air mixture occurs. The local region may be very small, as in the case of an electrically induced spark, a static spark, or a minute flame. The important point is that at this very small point in space, a temperature in excess of the ignition temperature has occurred in the presence of appropriate fuel and air (or oxygen). Such circumstances are not unusual, but from the investigative standpoint, it must always be accepted that they constitute a *minimum requirement for any fire whatever to result*. Many times, no source of such relatively high temperatures appears to be present, and the origin of a fire seems very mysterious. In such instances, it should be remembered that local high temperature is actually not difficult to achieve in a very small area. Fires have been kindled by rubbing two sticks, the flint and steel which strikes a very small, but hot, spark, and other similar means. In starting the fire, it is only the temperature that counts—not the area over which the temperature holds. A nail in a shoe heel, striking a rock, may kindle a fire, not because the shoe is hot, but only because an almost infinitesimally small region (the spark) reaches a very high temperature.

It would be rare indeed for any large amount of fuel to be heated past its ignition temperature. If any small portion of it is so heated, and the general temperature is above the flash point, a fire may result. This is why so many fires occur under conditions that do not seem to be hazardous in any serious degree. Some ignition temperatures of common fuels are indicated in Table 2.

TABLE 2. SPONTANEOUS IGNITION TEMPERATURES OF REPRESENTATIVE FUELS*

Substance	Temperature
	°F
Acetone	1000
Acetylene	635
Ammonia (anhydrous)	1204
Amyl acetate-n	714
Amyl alcohol	621
Aniline	1418
Benzene	1076
Benzine (petroleum ether)	475
Butane-n	806
Butane-iso	1010
Butyl alcohol-n	693
Butyl cellosolve	472
Camphor	871
Carbon disulfide	257
Carbon monoxide	1204
Castor oil	840
Cellosolve	460
Denatured alcohol (95%)	750
Dichloroethylene	856
Diethylcellosolve	406
Diethylether	366

*Data taken from "Properties of Flammable Liquids, Gases and Solids," *Loss Prevention Bulletin* No. 36.10, Factory Mutual Engineering Division, Boston, 1950.

(Continued)

TABLE 2. SPONTANEOUS IGNITION TEMPERATURES OF REPRESENTATIVE FUELS (Continued)

Substance	Temperature
	°F
Ethane	950
Ethyl acetate	907
Ethyl alcohol	799
Ethylene	1009
Ethylene dichloride	775
Ethylene glycol	775
Formaldehyde (gas)	806
Fuel oil #1	490
Fuel oil #2	494
Fuel oil #3	498
Fuel oil #4	505
Fuel oil #6	765
Gas, coal gas	1200
Gas, natural	(see methane gas)
Gas, oil gas	637
Gasoline	495
Glycerine	739
Hexane-*n*	477
Hydrogen	1076
Hydrogen sulfide	500
Kerosene (fuel oil #1)	490
Linseed oil	820
Lubricating oil (mineral)	500–700
Methane	999
Methyl alcohol	878
Methyl cellosolve	551
Methyl ethyl ketone	960
Mineral spirits	475
Naphtha (coal tar)	900–960
Naphtha (Stoddard solvent)	450–500
Naphtha (V.M. and P.)	450–500
Nitroglycerine	518
Octane-*n*	450
Petroleum ether	475
Phosphorus (red)	500

TABLE 2. SPONTANEOUS IGNITION
TEMPERATURES OF
REPRESENTATIVE FUELS
(Continued)

Substance	Temperature
	°F
Phosphorus (yellow)	86
Pine tar	671
Propane	874
Propyl alcohol-iso	852
Pyridine	1065
Resin oil	648
Soya bean oil	833
Sulfur	450
Toluene	1026
Tung oil	855
Turpentine	488

EXPLOSIVE LIMITS

Mixtures of flammable vapors or gases with air will explode only when they are within particular rather restricted ranges of composition. Some materials, such as hydrogen and carbon disulfide, have very wide ranges within which flame will propagate to give an explosion. However, most materials, such as hydrocarbons, show quite narrow limits of composition. Within a confined system, if the mixture will not explode, it also will not ignite. For this reason, the explosive limits are of the greatest significance in the interpretation of fires.

In the open, there are other variables that may alter the above statement. Assume that a mixture of vapor and air is too lean for an explosion; it may be that the efforts to ignite it will raise the temperature and increase the amount of material volatilized until a fire can result. More important, probably, is the too rich mixture which will not ignite. The application of a brief spark will not be expected to cause a fire. However, prolonged application of a flame will produce so much turbulence as to mix in additional air, and thus start a fire which with an overly rich mixture initially is expected to develop into what may be known as a "rolling fire" of high intensity. It must be emphasized that the reason for this effect is, in both cases, the alteration of the system by means of the heat applied by the ignition source. Such an alteration would not be possible in an enclosed system, and would not be effective in producing a fire any more than in producing an explosion. Thus, on a hot day, when the gasoline-air mixture inside a tank of gasoline is richer in gasoline vapor than the upper explosive limits, a spark could safely be produced inside the tank with no effect. At lower temperatures, when the vapors are allowed to form a mixture with the air that is within the explosive range, an explosion will surely result.

Some other conclusions of interest are related to the admixture of the vaporous fuel with air. A lean mixture (excess air) tends to produce a sharp explosion, in which all fuel is totally combusted, and often no fire will follow it. A rich mixture, on the other hand, results in a less vigorous explosion with less mechanical damage, producing soot, and usually is followed by additional fire since excess fuel is present. Thus, the heat effects and blackening are greater and the mechanical damage less when the mixture is rich than when it is lean. These points are of significance to the investigator and should be carefully noted.

Table 3 lists the explosive limits of representative materials.

TABLE 3. EXPLOSIVE LIMITS % BY VOLUME OF VAPOR AIR MIXTURES[*]

Substance	Lower	Upper
Acetone	2.15	13.0
Acetylene	2.5	80
Amyl acetate-*n*	1.1	–
Amyl alcohol-*n*	1.2	–
Benzene	1.4	8
Benzine (petroleum ether)	1.1	4.8
Butane-*n*	1.6	6.5
Butane-iso	1.6	8.5
Butyl alcohol-*n*	1.7	–
Carbon disulfide	1.0	50
Carbon monoxide	12.5	74.2
Cellosolve	2.6	15.7
Dichloroethylene, 1,1	5.6	11.4
Dichloroethylene, 1,2	9.7	12.8
Diethyl either	1.7	48.0
Ethane	3.3	10.6
Ethyl acetate	2.18	11.5
Ethyl alcohol	3.28	19
Ethylene	3.02	34
Ethylene dichloride	6.2	15.9
Ethylene glycol	3.2	–

[*]Data taken from: "Properties of Flammable Liquids, Gases and Solids," *Loss Prevention Bulletin* No. 36.10, Factory Mutual Engineering Division, Boston, 1950.

TABLE 3. EXPLOSIVE LIMITS % BY VOLUME OF VAPOR AIR MIXTURES (Continued)

Substance	Lower	Upper
Gas, coal gas	5.3	31
Gas, natural	4.8	13.5
Gas, oil gas	6.0	13.5
Gasoline	1.3	6
Hexane-*n*	1.25	6.9
Hydrogen	4.1	74.2
Hydrogen sulfide	4.3	45.5
Kerosene (fuel oil #1)	1.16	6.0
Methane	5.3	13.9
Methyl alcohol	6.0	36.5
Methyl ethyl ketone	1.81	11.5
Naphtha (Stoddard solvent)	1.1	6.0
Naphtha (V.M. and P.)	0.92	6.0
Octane-*n*	0.84	3.2
Petroleum ether	1.4	5.9
Propane	2.3	7.3
Propyl alcohol-iso	2.5	—
Pyridine	1.8	12.4
Toluene	1.27	7.0
Turpentine	0.8	—

BOILING POINTS

Boiling points are related to fire investigation. They are, however, generally of secondary significance as compared with other properties discussed in this chapter. Most liquids may be expected to be heated above their boiling points in a fire. However, the very important distinction between a liquid fuel that is reasonably pure, and will therefore have a reasonably definite boiling point, and one that is a mixture of many components of variable boiling points must be grasped and applied to the investigation.

Table 4 lists a considerable number of pure liquid fuels. Each of these will have a rather definite boiling point at which an entire sample may be distilled without a change in temperature of the vapors. Only when the fuel is pure can these figures be taken at their face value and applied without further consideration. The general public has available to it relatively few pure liquid fuels as compared with the much larger number of mixtures which have a boiling range rather than a boiling point. Such mixtures range from crude oil, with perhaps hundreds of individual compounds, to a large number of manufactured mixtures for special purposes such as cleaning, industrial solvent action, etc. When heated, such mixtures allow the distillation of their most volatile constituents first, followed in turn by constituents of lesser and lesser volatility. In general, it is the most volatile constituent of the mixture that is of most significance as it relates to the fire hazard.

In initiating a fire, materials of high boiling point will rarely be above their flash points at ambient temperatures, and only the more volatile (low boiling point) will be expected to show special hazard. As a rule the boiling points and the flash points tend to parallel each other, and the low boiling material will usually have a low flash point as well. If it exists in a mixture, this does not greatly diminish its intrinsic danger as a source of fire hazard.

The boiling points of a limited number of materials of interest are given in Table 4.

TABLE 4. BOILING POINTS OF SOME REPRESENTATIVE MATERIALS*

Substance	Boiling Point °F
Acetone	134
Amyl acetate	300
Benzene	176
Benzine (petroleum ether)	100–160
Butane-n	33
Butyl acetate	260
Butyl alcohol-n	243
Butyl cellosolve	340
Carbon disulfide	114
Cellosolve	275
Denatured alcohol (95%)	175
Dichloroethylene, 1,2	141
Diethyl ether	95
Ethane	−128
Gasoline	100–400
Methyl alcohol	147
Methyl ethyl ketone	176
Mineral spirits	300
Naphtha (Stoddard solvent)	300–400
Naphtha (V.M. and P.)	212–320

*Data taken from: "Properties of Flammable Liquids, Gases and Solids," *Loss Prevention Bulletin* No. 36.10, Factory Mutual Engineering Division, Boston, 1950.

TABLE 4. BOILING POINTS OF SOME
REPRESENTATIVE MATERIALS
(Continued)

Substance	Boiling Point °F
Octane-n	257
Petroleum ether	100–160
Propane	–45
Toluene	232
Turpentine	300

VAPOR DENSITY

Vapor density is important in the interpretation of fires or explosions in which vapors are involved, but ordinarily only with regard to their weight and density in relation to the weight and density of air. Naturally, all vapors heavier than air tend to drop through the air into which they are released until they encounter an obstruction, such as a floor, after which they tend to spread at this level, much in the same manner as if they were liquids. On the other hand, vapors lighter than air tend to rise through the air until an obstruction, such as a ceiling, is encountered, after which they spread at the high level. The tendency is less absolute than with liquids, because the difference in density as compared with air is much less than with any liquid, and air currents produce mixing. Some mixing with the air is inevitable, and the nearer to air the density is, the greater the mixing. This effect is due to the property known as *diffusion*. All gases, regardless of their molecular weight or vapor density, will tend to diffuse into each other when the opportunity exists. The difference between them in this regard is that when the fuel vapor is much heavier than air, its diffusion rate will be less than when it is closer to that of the air with which it mutually diffuses. It is interesting to compare the lighter hydrocarbons in this regard.

Methane, the lightest, has already been noted as being lighter than air. Thus, it tends to rise in a room filled with air, but also being lighter, its tendency to diffuse is greater than heavier molecules, and the separation based on difference of vapor density is less perfect than heavy fuel vapors. *Ethane*, next in the series, has a molecular weight, and therefore a vapor density almost exactly equal to that of air. Thus, there exists no gravitational separation and free diffusion tends to produce uniform admixture rather rapidly. Ethane is generally a minor component of fuels, and therefore of less importance than either methane or the heavier gases. *Propane*, next heavier in the series, will tend to settle through air to a significant degree, but because it also is rather light, diffusion will cause its admixture with the air rather rapidly. *Butane*, fourth in the series, is the first material of the series that is heavy enough to reduce greatly the rate of diffusion

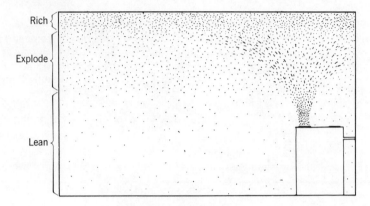

Figure 1(a). *Natural gas in a room.*

Figure 1(b). *Gasoline in a room.*

into air, and thus slow the process of developing explosive mixtures. All higher hydrocarbons, characteristic of most petroleum products will, of course, have slower diffusion rates along with their greater vapor density.

The effect of releasing vapors, both lighter (natural gas) and heavier (gasoline) than air, into an air-filled region is shown in the figure. It will be apparent that various concentration layers will exist. Some of these will be within explosive (flammable) limits, while there may be others in which the mixture is either too lean or too rich to be ignited by any means.

In the process of diffusion and admixture of flammable gases and vapors with air, the effects of temperature and of air currents must not be overlooked. The simplified situation illustrated is one which tends to occur in a closed room in which there is little or no physical activity, and in which the gases are at the same temperatures. Even butane, if heated sufficiently, will rise to the top of the room much like methane, because the vapor density drops rapidly with increased temperature. When the gas mixture is formed from gases at different temperatures, the temperature alone may control their movement initially, until the temperature difference is equalized.

Movement of the gases, produced by physical activity such as persons moving about, fans, machinery, and similar causes, will mix them much more rapidly than in a static situation. For example, if gasoline is leaking from a carburetor of a running motor, the wind produced by the fan will rapidly dissipate the gasoline fumes and produce highly hazardous conditions. As a rule, the alert investigator will note such conditions as those mentioned, and make allowance for them in his consideration of the fire's cause and sequence.

Nearly all flammable vapors and gases are heavier than air. In the vapor or gaseous state, the relative density can be calculated simply by comparing the molecular weight of the vaporous substance with the mean molecular weight of air, which is approximately 29. The lightest gas, and a highly flammable one, is hydrogen, which has a molecular weight of only 2. Methane, the chief constituent of natural gas, is also lighter than air, in the proportion of 16 to 29. Ethane, rarely encountered except as an admixture with other petroleum products, has a molecular weight of 30, making it almost exactly the same weight or density as air. It will therefore easily mix with air without any significant separation to lower or higher levels. All other important hydrocarbons except acetylene and, in fact, virtually all other flammable vapors are heavier than air and will tend to sink at variable rates when released into air. Octane, a major constituent of gasoline, for example, has a molecular weight of 114, which makes it almost exactly four times as heavy as air. Gasoline fumes sink rapidly in still air to floor or ground surface. They will, therefore, pour into low regions or down drains and will carry their intrinsic hazard to the lowest possible region. Explosions of gasoline fumes correspondingly tend to blow open the lower portions of structures such as walls. Natural gas, on the other hand, tends to attack the upper portions of such structures, although the admixture with air will be more rapid and the forces therefore more widespread. While under ideal conditions, no selective damage to the lower portions of confining structures is expected, in practice this is not always the case. Because of the rapid diffusion of natural gas in air, the vertical distribution may not be marked as was illustrated above and forces exerted by the explosion of the mixture may be quite uniformly distributed. Another factor of great consequence here is the relative structural integrity of various portions of the structure. It is frequently noted that in natural gas explosions, it is the bottom of the wall that moves rather than the top as would be expected. This can result from the fact that the bottom may not be as well anchored to the foundation as the top is to the structure above. Thus, too great reliance on the general principle can, in such an instance, lead to an incorrect conclusion.

HEAT OUTPUT

Another property of fuels that is of only occasional importance to the investigator is the heat output of combustion, or more simply the *heat of combustion*. Its value to the investigator lies chiefly in the comparison of intensity of burn with the possibilities of various fuels that may have been involved. Some of the most common fuels, such as wood, are so variable as to give only approximate notions of the actual output in a specific instance. The following table summarizes the heat of combustion of a few of the most important fuels from the standpoint of fire investigation.

Material	Heat of Combustion
Crude oil	19,650 BTU/gal.
Diesel fuel	19,550 BTU/lb.
Gasoline	19,250 BTU/lb.
Methane	995 BTU/cu. ft.
Natural gas	128-1868 BTU/cu. ft.
Octane	121,300 BTU/gal.
Wood	7,500 BTU/lb.
Coal—Bituminous	11,000-14,000 BTU/lb.
Anthracite	13,351 BTU/lb.

BTU (British Thermal Unit) is defined as the quantity of heat required to raise the temperature of 1 pound of water $1°F$. at or near its point of maximum density.

HYDROCARBON FUELS

By far, the most important fuels from the standpoint of fire investigation are the hydrocarbons, which range from the light gas methane to higher molecular weight compounds in several homologous series. Because of their unique importance, some of their properties will be discussed.

Natural gas, consisting chiefly of methane (80–95 percent) with smaller quantities of ethane, propane, and butane, is of obvious significance. Natural gas from different areas shows considerable variation in composition and may also include some incombustible gases. For most purposes, the investigator may consider its properties to be the same as those of methane without introducing serious error.

Petroleum in its crude state is a thick oil varying in color from a light brown to black. It contains a very large number of compounds and different types of compounds, which are separated to a considerable extent in the manufacture of petroleum fuels. Products obtained from the distillation of petroleum include petroleum ether, gasoline, kerosene, and other middle oils including naphtha type mixtures, heavy oils, vaseline, paraffin, wax, and coke. There are four principal types of crude oil: Pennsylvania (containing paraffin compounds, C_nH_{2n+2}); California (containing naphthene, C_nH_{2n}, and some benzene and its homologues); Texas (containing hydrocarbons of the C_nH_{2n-2} and C_nH_{2n-4} series); and Baku (containing naphthene). The composition of these petroleum types is indicated in Table 5.

TABLE 5. COMPOSITION OF CRUDE OILS

	C	H	N*	S	O	Sp. gr.
California	84.43%	10.99%	0.65%	0.59%	3.34%	0.962%
Texas	84.60	10.90		1.63	2.87	
Java	87.1	12.0				
Baku	85.3	11.6				0.923

*Varies $0.25 - 1.0\%$.

Gasoline, perhaps the most important fuel of petroleum origin, is a mixture of volatile, low-boiling hydrocarbons. There are two main types of gasoline: straight run and cracked. The average molecular weight of gasoline is usually taken as that of octane, about 114. The composition of various gasolines, concerning type of compound, is shown in Table 6.

TABLE 6. COMPOSITION OF GASOLINES

	Paraffins	Naphthenes	Olefins	Aromatics
	Straight Run			
Mid-Continent	72.9%	22.0%	1.9%	3.2%
Mixed California	58.9	31.6	2.2	7.3
Pennsylvania	82.5	15.3	2.1	Trace
Mexico	82.3	10.9	1.5	5.3
Venezuela	71.0	20.4	0	8.6
Michigan	85.2	7.4	2.9	4.5
	Cracked			
Pennsylvania Low temp.	65.8%	6.0%	11.6%	17.4%
Mirando Gulf Coast	33.1	30.8	13.4	22.7

(Continued)

TABLE 6. COMPOSITION OF GASOLINES *(Continued)*

	Paraffins	Naphthenes	Olefins	Aromatics
		Cracked		
High Temp. Vapor phase Venezuela	13.5	11.5	34.1	40.9
Mixed Mid-Continent Low temp.	47.0	11.0	17.0	25.0

As an indication of the comparative complexity of gasoline, as well as the variations which are possible in its composition, the compounds that may exist in it are shown in the following table. It must be realized that some of these compounds, and therefore variability in the finished product, relate directly to the source of the crude, but this is not necessarily invariable. Cracked gasolines are prepared by breaking up larger molecules into the smaller ones that are better suited to the purposes of a volatile fuel. Here also a wide variety of compounds may be formed.

Paraffins: (C_nH_{2n+2})		Naphthenes: C_nH_{2n} (ring forms)	
N-pentane	C_5H_{12}	Cyclo-pentane pentamethylene	C_5H_{10}
Iso-pentane	C_5H_{12}	Methyl-penta-methylene	C_6H_{12}
N-hexane	C_6H_{14}	Methyl-cyclo-pentane	C_6H_{12}
Iso-hexane	C_6H_{14}	Cyclo-hexane	C_6H_{12}
N-heptane	C_7H_{16}	Hexa-methylene	C_6H_{12}
Iso-heptane	C_7H_{16}	Hexa-hydro-benzene	C_6H_{12}
N-octane	C_8H_{18}	Methyl-cyclo-hexane	C_7H_{14}
Iso-octane	C_8H_{18}	Methyl-hexa-methylene	C_7H_{14}
N-nonane	C_9H_{20}	Hexa-hydro-toluene	C_7H_{14}

Olefins: C_nH_{2n} (chain forms)		Aromatics: C_nH_{2n-6}	
Ethylene	C_2H_4	Benzene	C_6H_6
Propylene	C_3H_6	Toluene	C_7H_8
Butylene	C_4H_8	Xylene	C_8H_{10}
		Cumene	C_9H_{12}

Kerosene, and the distillate fuels made from more aromatic stocks and of higher volatility were minor in their role as fuel until the Diesel motor became a common device. Although diesel fuel is similar to kerosene, it may be modified in various ways. The modern jet engine and other turbine drive engines further increased the general utilization and significance of kerosene-like fuels. This is because they have considerably higher BTU values than the more volatile materials and therefore hold a weight premium as related to the power obtainable from them. At the same time, kerosene and similar hydrocarbon mixtures have long been of critical significance in the setting of deliberate fires. Their lower volatility presents less hazard to the user. The liquid is more persistent, so that less haste is required in the ignition as well as less danger of explosion. Their compositions are, like those of the gasolines, complex, but involve several series of higher molecular weight compounds.

NON-HYDROCARBON LIQUID FUELS

In the tables presented above, there were a great many materials other than hydrocarbons listed. Some of them will have considerable significance to the fire investigator because they can be used as accelerants, and can create conditions predisposing to accidental fire as well. Such materials as the various alcohols are widely available, often for special purposes, such as denatured ethyl alcohol used as a painter's solvent. Ketones such as acetone and methyl, ethyl ketone are also widely used as solvent components. Turpentine is often employed for thinning paints. All of these and many other such materials fall in the category defined above. They require mention primarily because a predisposition to think of liquid fuels only in terms of hydrocarbons can be totally unrealistic in a specific instance. Industrial fires especially are often fed and intensified to extremes by special liquids used for solvents and other industrial purposes. Inasmuch as their general properties have already been listed, it is unnecessary to attempt individual treatment of them, especially in view of their very large number and widely different compositions and properties. Special considerations of some of these will appear from point to point in this volume.

Supplemental References

Coward, H. F., and Jones, G. W. *Limits of Flammability of Gases and Vapors*. Bureau of Mines Bull. 503, 1952.

Matson, A. F. and Dufour, R. E. "The Lower Limit of Flammability and the Autogenous Ignition Temperature of Certain Common Solvent Vapors Encountered in Ovens." *Underwriters' Laboratories Bull. Of Research*, No. 43, Jan. 1953.

Nuckolls, A. H. "Classification of the Hazards of Liquids." *Underwriters' Laboratories Bull. of Research*, No. 29, Dec. 1945.

Zabetakis, M. G., Furno, A. L., and Jones, G. W. "Minimum Spontaneous Ignition Temperatures of Combustibles in Air." *Ind. & Eng. Chem.*, 46, 2173, 1954.

5

Combustion Properties of Solid Fuels

In addition to the liquids and vapors for which extensive physical and chemical properties have been listed, there is a large group of very common fuels for which accurate and precise data cannot generally be tabulated. Many of the typical combustion properties may have little or no meaning for this group. For this reason, it is not often possible to list such properties as flash point or explosive range for the most common fuels of all: wood, coal, paper, and the like. Despite this limitation, the properties of these fuels are of the greatest importance, and to some extent they can be indicated, either in numerical data or in general terms.

The chief reason that any numerical values at all can be attributed to the solid fuels is that when heated most, if not all of them, undergo heat decomposition or pyrolysis, with production of simpler molecular species that do have definite and known properties. The matter of pyrolysis is dealt with in a following chapter which will elucidate the nature of the effect being discussed. However, here also, precise values are often not available for two main reasons:

1. In some instances the pyrolysis of solid materials has been inadequately studied or not studied at all.
2. The pyrolysis of a single solid material gives rise to a great many simpler products in a mixture, so that the properties are not those of pure materials.

This is not to say that research work has not been done. However, much of it has been directed either at basic mechanisms not involved in investigation, or more commonly to industrial application of the fuels in question. Much investigation has also been carried out in connection with such matters as mine explosions which will rarely be of interest to the general fire investigator.

COAL

From the standpoint of its general use, coal ranks very high as a widely used fuel in industrial and general heating purposes. It is of secondary consequence as a problem for the fire investigator. It is an organic mixture made up of approximately 80 percent carbon, 5 percent hydrogen, and 12 percent oxygen, nitrogen, and sulfur. There are two major types of coal: (1) bituminous or soft coal and (2) anthracite or hard coal. Smaller amounts of coal in an intermediate range also occur. The average elemental content of bituminous coal is 80 percent carbon, 5.5 percent hydrogen, 0.9 percent sulfur, 1.5 percent nitrogen, 5.0 percent oxygen, and 7.0 percent mineral ash. Anthracite contains 91.29 percent carbon, 2.9 percent hydrogen, 2.75 percent oxygen and nitrogen, and 3.06 percent mineral ash. The nitrogen content varies from 0.58 percent to 2.85 percent while sulfur varies from 0.63 percent to 1.0 percent.

There is progressively less use of coal for heating the interiors of structures as compared with heating practices of the past. It continues to be of considerable industrial importance as a fuel, and coal mining carries with it hazards that relate to the combustion properties of the coal involved. For example, both stored and unmined coal is subject, under proper conditions, to *spontaneous combustion* (1).* Moisture is essential to this occurrence, and the amount of moisture must be within suitable ranges, as is typical of spontaneous combustion with other materials, discussed elsewhere.

---

*Numbers in parentheses refer to references at end of chapter.

Johnson and Auth (2) present useful information on properties of coal combustion. They list the combustion temperature of bituminous coal as 766° F. and that of anthracite as 925° F. Flame temperatures are listed as follows:

Dark red	975° F.
Dull red	1290° F.
Dull cherry red	1470° F.
Full cherry red	1650° F.
Clear cherry red	1830° F.
Deep orange	2010° F.
White	2370° F.
Bright white	2550° F.
Dazzling white	2730° F.

It will have been noted that coal is highly carbonaceous. It also contains a number of other elements which proves the existence of a variety of compounds which are subject to pyrolysis with formation of simpler compounds. Coal differs from most organic solid fuels in that the quantity of such compounds is relatively small. The more there are present, the more easily it is for the material to ignite. Coal requires considerable input of heat from kindling sources to initiate a self sustaining combustion because of the small quantity of pyrolyzable compounds.

WOOD

Far more wood is burned as fuel in structural and outdoor fires than any other solid material. Thus, its properties as fuel, and as regards its behavior during combustion, are of greater importance in general than those of any other solid combustible material.

The term "wood" is generic and covers a wide variety of materials of vegetable origin, the chief component of which is cellulose. In addition, numerous other materials are present in wood including the hemicelluloses, lignin, resins, salts, and water. The chemical make-up of cellulose has been discussed previously (Chapter 2). Wood products, which are materials composed chiefly of cellulose, include all kinds of manufactured boards and panels, formed wood items such as furniture, and a host of paper and cardboard products manufactured from wood pulp for the most part.

Wood is obtained from several varieties of trees, some resinous, others not; some dense, others diffuse in structure. These also include a wide variety of water content, volatile material obtained at raised temperatures, and other complicating factors. In addition to their diverse origins, wooden materials are also greatly altered by manufacturing processes, with a variety of prepared sheet and board materials, and with all types of treatment and finish.

Some woods, e.g., pine, will ignite readily and burn vigorously because they contain resinous materials that provide volatile flammable vapors when heated. These serve to add greatly to the limited ability of cellulose and lignin to support the fire. The effect is not uniform with soft woods. Some are relatively resistant to fire, as discussed below in connection with redwood.

As a rule, hardwoods, e.g., oak, are difficult to ignite but are capable of generating much heat and extended combustion when burning. Thus, they present a lessened fire hazard as compared to the more combustible soft woods but create a hotter and more protracted fire. Ease of ignition of wood is correlated chiefly with the content of materials, e.g., pitch, that readily decompose to liberate combustible gases. The soft woods which do not have such materials and the hard woods, are more alike in their ease of ignition than they are to the resinous softwoods. An extreme case of poorly ignitable softwood is redwood, which is almost incapable of producing a self-sustaining fire, and far more difficult to burn than the hardwoods. The reason for this resistance may not have been related to pitch content. Certainly the charred redwood is adherent to the unburned wood below and serves to insulate it from flame, rather than breaking and flaking away as many woods do when charred.

The great resistance of redwood to fire was demonstrated by an experiment. A four-inch thick wall of redwood timber was exposed to the direct flames of a gasoline burner, 18 inches in length, for one hour and 20 minutes. At the end of this period, the burner was removed. Little flame was emitted from the redwood, and the glowing charred wood shortly cooled and extinguished. The penetration was less than one inch except in a very small region in which a few flames persisted. Nearly five gallons of gasoline were burned, and the flames were as much as four to five feet in height during most of this time.

Other than the extraneous components of some woods, their main fuel is cellulose. This is also the material of which cotton is composed, and is the major constituent of most papers. When finely divided, it is obviously very flammable, but with a low output of heat. Cellulose is a derivative of sugar which already is highly oxidized, so that the molecule has proportionately little fuel to provide for oxidation. Ease of ignition makes it suitable for kindling of fire. However, the total heat output is small as compared with more reduced fuels such as coal and hydrocarbon fuels. Thus, fires that have only wood for fuel tend to be far less intense than those that are fed by organic liquids such as are often used as fire accelerants.

While wood is not highly carbonaceous, being partially oxidized, one of its main combustion products, *charcoal*, is. There are few, if any, fuels higher in carbon content than charcoal; thus, charcoal is an excellent fuel for high heat production. It is correspondingly difficult to ignite, and, lacking volatile materials, gives little flame. The small blue flames associated with the burning of charcoal are due to formation of carbon monoxide gas which burns rapidly close to the surface of the charcoal from which it was generated. The charcoal fire is largely a smouldering fire, that is, one without flame, but intensely hot. Here, the hot solid is reacting with gaseous oxygen to produce combustion on and slightly below the surface. Charcoal is, of course, formed by pyrolyzing and destructively distilling the volatile materials out of the wood, leaving behind the nonvolatile constituents of the wood, chiefly carbon.

Coke, formed similarly from coal or petroleum, shows burning characteristics much like those of charcoal, and for similar reasons.

Combustion properties of wood have been studied extensively, both because of variable fire risk and for purposes of interpretation of the fire after it has occurred. Browning (3) considered extensively the ignition temperature of wood and its behavior when heated. He states: "The ignition temperature usually defined in comparing fire resistance of woods is that at which wood gives off volatile products which can be ignited from an external source of higher temperature. . . . . It may be as low as 228° C. (442° F.) although results are variable and dependent on conditions of testing." He also points out that the migration of inorganic materials to the surface during drying reduces the surface flammability.

Another term of importance is the *autogenous or self-ignition temperature*. This varies from the ignition temperature as defined above in that it is the lowest initial temperature from which, under favorable conditions, a material will heat spontaneously until glow or flame results. This temperature may be related closely to the ignition temperature of some pyrolysis products formed from the wood; but it involves other factors as well. It is highly dependent on the size of

TABLE 1. EFFECT OF TEMPERATURE AND TIME OF EXPOSURE UPON THE IGNITION OF WOOD (4)

Temp.		Duration of Exposure before Ignition (min.)								
°C	°F	Long Leaf	Red Oak	Tamarack	Western Larch	Noble Fir	Eastern Hemlock	Red-wood	Sitka Spruce	Basswood
180	356	14.3	20.0	29.9	30.8	–	–	28.5	40.0	–
200	392	11.8	13.3	14.5	25.0	–	13.3	18.5	19.6	14.5
225	437	8.7	8.1	9.0	17.0	15.8	7.2	10.4	8.3	9.6
250	482	6.0	4.7	6.0	9.5	9.3	4.0	6.0	5.3	6.0
300	572	2.3	1.6	2.3	3.5	2.3	2.2	1.9	2.1	1.6
350	662	1.4	1.2	.8	1.5	1.2	1.2	.8	1.0	1.2
430	806	.5	.5	.5	.5	.3	.3	.3	.3	.3
Reported Specific Gravity		.70	.68	.60	.48	.46	.46	.35	.34	.31

the sample, the testing apparatus used, the rate of heating and other factors. A major one is the temperature at which the oxidative reaction becomes exothermic and, therefore, self-sustaining. This temperature appears to be about 270° to 280°C. (518° to 536°F.), but ignition temperatures ranging from 190° to 555°C. (374° to 1031°F.) have been reported. Useful compilations have been published by McNaughton (4) and by Graf (5), as shown in Tables 1 and 2.

TABLE 2.   IGNITION TEMPERATURES OF WOODS (5)

	Class	Ign. Temp., °F.	Wt. of Sample, Grams	Rate of Temp. Rise/ hr. °F.	Air flow, cu.ft./min.	Moisture, % of wt.	Sample Size
Tan bark oak	Hard	448	12	15	0.05	Less than 7%	1 X 2¼
Ponderosa pine (sapwood)	Soft	457	10	4	0.05	7%	1 X 2¼
Ponderosa pine Heart	Soft	500	8	24	0.05	Less than 7%	1 X 2¼
Redwood	Soft	467	8	18	0.05	7%	1 X 2¼
Western red cedar	Soft	468	7	17	0.05	Less than 7%	1 X 2¼
Sitka spruce	Soft	482	10	16	0.05	7%	1 X 2¼
Douglas fir	Soft	489	10	16	0.05	7%	1 X 2¼
Oregon oak	Hard	500	10	30	0.05	7%	1 X 2¼

Angell (6) lists ignition temperatures as follows: sound southern pine, 400°F.; sound redwood, 400°F.; decayed southern pine, 300°F. Chemical fireproofing increased these values to greater than 1200°F.

A variation in the susceptibility of wood to ignition and rapidity of combustion is provided in some of the manufactured wood products so extensively utilized in modern building. Some of these require special consideration, because they can modify the usual burning characteristics of the wood from which they are derived. *Plywood* and *veneer board* have become common items of use by builders for special purposes. Both are manufactured from very thin sheets of wood, laminated to each other by layers of adhesive. Veneer board generally is made with a top layer of clear hardwood, laminated to less expensive base layers. It is used almost entirely for interior finish, although some types may be used on the exterior.

It must be remembered that the finer or thinner wood is cut, the better it ignites and the faster it burns because of the exposure of a larger surface to heat and air. Any board of this type will burn in a manner generally like that of a solid wood board of the same dimensions, except for the influence, if any, of the adhesive or coating. However, if the adhesive softens under heat, the layers will delaminate and open like the pages of a book. This will expose a greatly increased surface area to the flames and cause comparatively fiercer burning. Thus, in this regard, the adhesive is far more important than other considerations. Plyboard from the most common sources is not much different in its combustion properties than other wood of the same thickness. Some imported veneer boards are made with unsuitable adhesive, and may greatly accelerate a fire involving them.

Another factor that may contribute to the influence of these materials is the combustion properties of the adhesive itself. Often the adhesive is less readily combustible than the wood layers; if it is more combustible, it may contribute what is effectively a fire accelerant. In the evaluation of any fire involving these materials, special attention to the adhesives involved is essential.

This author has investigated a number of fires in which plyboard and veneer board have been heavily involved. In no instance has the origin of the fire been traced to such materials, but the rapidity of buildup and spread of the fire was nearly always markedly increased. In one instance of a fire with an electrical origin, an extremely thin plywood finish at some distance from the origin had produced such severe local burning that the location was for a time suspected as the point of origin. This illustrates the necessity of checking on all possible types of origin and causation before adopting any one of them. Failure to do this has led to numerous errors on the part of some fire investigators.

Other cellulosic building materials, not necessarily made from wood, but alike in their combustion properties, include Cellotex and similar boards made from compressed and bonded fibrous materials of a woody nature. These are used as acoustical tiles to cover ceilings, and in larger sheets as insulation or for other such purposes. Tests show that most of these materials are relatively fire resistant, although subject to a smoldering fire in their interiors on occasion. In a large surrounding fire, they burn well and contribute much fuel. Thus, it is common to see that in a room fire, all of the acoustical tiles have fallen and burned on the floor. They fall readily because they are generally glued into place and the heat softens the adhesive. In this way, they contribute significantly to the intensity of the fire, but only rarely are they concerned in starting it. Once on fire, they are difficult to extinguish, but tend to continue smouldering.

PAPER

Paper is one of the more interesting substances involved in fires, partially because of its frequent use in kindling fires and partially because of its unusual properties as regards combustion. Everyone who has kindled a fire with paper will realize that newspaper works well, while a picture magazine makes very poor kindling. The reasons are simple but not always understood.

The basic ingredient of all paper is cellulose, the same material as that in cotton and a major constituent of wood, from both of which paper is made. While cellulose is readily combustible, not all of the extraneous constituents of paper are. "Slick" papers have a very high content of clay, which is not combustible. Many writing papers also have a high clay content. More clay goes into the manufacture of paper than goes into manufacturing ceramics, which are based almost exclusively on their clay content. The differences in the combustion properties of various papers is readily demonstrated by burning a piece of newspaper, which leaves a small and very light ash, followed by burning a piece of the heavy, smooth, "slick" paper, which leaves a very heavy ash. The ash in most instances is the clay, although many specialty papers also contain such constituents as titania, barium sulfate, and other non-combustibles.

Graf (5) studied the ignition temperatures of a variety of papers. With few exceptions he found the ignition temperature to be between 425° and 475°F. At 300°F., most appeared normal; many tanned in the range up to 350°F., went from tan to brown by reaching about 400°F., and if not ignited, became brown to black at about 450°F.

Paper burns well only because it is thin, providing much surface to heat and air. When it is stacked, it no longer has much exposed surface, and a pile of stacked papers is virtually impossible to burn, although it will char on the surface. Ventilation is absent on the inside of such a stack, thus making paper a good insulator against penetration of heat from a surrounding fire. In fact, a fire may be smothered by paper quite effectively, merely by covering the burning surface with a few layers of it. The air supply is thereby shut off, and the fire is quenched. This is true even of newspaper, which is among the most combustible types of paper in existence. Firemen sometimes use this procedure to extinguish small fires.

Thus, in interpreting a fire, it is not the presence of paper that is significant but its distribution in terms of the exposed surface. Many times, the presence of a pile of papers is taken as evidence for the source of a greater fire. While possible, because paper is often used to kindle fire, its mere presence is not evidence of anything unless supported by additional facts. For example, the utilization of a pile of paper to which is added a liquid accelerant is not uncommon. In such an instance the accelerant burns at the surface of the pile, the paper serving as a wick to continue feeding the surface fire. The residue of such a pile of paper is a specially good place to look for traces of unburned accelerant. Here, the primary purpose of the paper is to hold the accelerant, not to contribute to the fire by its own combustion.

PLASTIC

Plastic, another material often involved in fire, has special properties that make it of great interest. There are many types of plastics. Some, such as Teflon (tetrafluoropolyethylene), are not combustible, and they show heat damage only at considerably raised temperatures. Others, such as nitrocellulose and celluloid, are violently flammable. Most of the commonly used plastics range between these two extremes. The polystyrenes, methacrylates, vinyls, cellulose acetates, and the polyethylenes are combustible but do not burn readily except in a hot surrounding fire, to which they may add great intensity of combustion. Many plastics melt before they burn, then decompose to form combustible vapors which feed the flames. At the stage of melting, and before thermal decomposition becomes significant, they pose little intrinsic hazard. Once they reach the stage of vigorous burning some of them are excellent fuel and difficult to extinguish.

The fire hazard associated with plastics, although it may be considerable with a few of them, tends to become exaggerated in the popular understanding. The reasons will be apparent in the following tables taken from the literature of plastics.

TABLE 3. PLASTICS BEHAVIOR ON HEATING (7)

Material	Treatment	Result
Formvar (polyvinyl formal)	Heated in test tube	Melts, some discoloration, slight charring, fishy odor, and smell of formaldehyde.
Alvar (polyvinyl acetal)	Heated in test tube	Melts, some discolor, slight charring, no smell of acetaldehyde, but this can be detected with Schiffs reagent. Slight fishy odor.
Polyvinyl chloride	Heated in test tube	Browns immediately becoming black. Little melting, copious evolution of HCl detected by smell and NH_3.
Mixed vinyl chloride acetate polymer	Heated in test tube	Like polyvinyl chloride.
Polystyrene	Heated in test tube	Melts to clear liquid which boils. Very slight discolor, charring. Smell of monomer.
Methyl methacrylate	Heated in test tube	Does not melt or char appreciably, decomposes and monomer distills off.

TABLE 3.   PLASTICS BEHAVIOR ON HEATING (7) (Continued)

Material	Treatment	Result
Bakelite	Heated in tube and flame	Presence of wood flour causes much charring and evolution of smoke. This disguises any characteristic odor. Without wood-flour, phenol and formaldehyde can be detected.
Urea-formaldehyde	Heated in flame	Strong smell of formaldehyde and NH_3. Much charring but highly non-inflammable.
Urea-formaldehyde	Heated in tube	Little smell of formaldehyde. Principal odor that of ammonia and pyridine.
Thiourea-formaldehyde	Heated in tube	Pronounced smell of H_2S and NH_3.
Casein	Heated in flame	Chars readily, pronounced smell of burning protein. More flammable than urea-formaldehyde. Smell very pronounced and disgusting.
Cellulose acetate	Heated in flame	Melts and chars, very pungent smell of burning cellulose and HAc.
Ethyl cellulose	Heated in flame	Chars, readily melts, smell of burning cellulose together with unpleasant oily smell.
Cellulose acetobutyrate	Heated in flame	Chars and melts. Distinct and characteristic smell of butyric acid.

The Modern Plastics Encyclopedia for 1964 (8) lists a useful table of burning rates, as shown in Table 4.

TABLE 4. BURNING RATES OF THERMOPLASTICS (8)

Thermoplastics	Burning Rate
Acetal polymer and copolymer	Slow
Acrylic	Slow
Cellulosic	
Ethyl cellulose	Slow
Cellulose acetate	Slow to self-extinguishing
Cellulose propionate	Slow
Cellulose acetate butyrate	Slow
Cellulose nitrate	Very fast
Chlorinated polyether	Self-extinguishing
Nylon	Self-extinguishing
Polyethylene	Very slow
Polypropylene	Slow to self-extinguishing
Polychlorotrifluoro-ethylene	None
Polystyrene	Slow
Unless glass filled	Fast to self-extinguishing
Polycarbonate	Non-burning to self-extinguishing
Phenoxy	Slow to self-extinguishing
Vinyl polymers and copolymers	Slow to self-extinguishing

Actual ignition temperatures, compiled by Lever and Rhys (9) and shown in Table 5, shows that, in general, plastics in common use ignite at temperatures considerably in excess of those of many liquid fuels and cellulosic products including wood and paper. They, therefore, constitute a minimum fire hazard as compared with such materials.

TABLE 5. IGNITION TEMPERATURE OF PLASTICS (9)

	Ignition Temperature, °F.	
	With Ignition Flame	Self-Ignition
Phenolic laminate	644	968
Phenolic cotton fabric laminate	642	918
Melamine wood flour filled	720	932
Melamine mineral filled	898	1,130
Polystyrene compression moulded	653	910
Polyvinyl-chloride-acetate, clear	680	1,112
Cellulose acetate	580	842
Polymethyl methacrylate	574	860

Plastic skylights, windows, and the like pose no special fire hazard. Their melting and final burning will only occur under conditions that would much earlier break glass used for similar purposes. If anything, their chief danger lies in the fact that because they do not break, and so ventilate the fire, they may force it to spread unduly within the confined area before a final breakthrough when the fire is drawn to the exit, and thereby somewhat localized.

PAINT

Paint is another product which is greatly misunderstood. In part this is due to the tremendous diversity of products, all designated as paint. Historically, paint consisted primarily of a drying oil, in which mineral pigments were suspended, and the mixture then thinned by turpentine or a mineral oil similar to gasoline or kerosene. The thinner was lost shortly after application of the paint, so it was dangerous only during the time of application. Drying oils are combustible, even after they are completely dry. Mineral pigments, on the other hand, are not combustible and offer some resistance to the spread of a fire. Thus, the local conditions are all important. In general, it can be said that paint tends to retard fire somewhat, but not enough to matter greatly in a major conflagration. It was also noted that when paint was applied in many layers, each on top of the other without intermediate removal of the old paint, a vigorous fire in the neighborhood would cause the paint to burn with considerable violence, and augment the fire. This danger was appreciated during the Second World War, when battleships which had been painted repeatedly were struck by bombs that produced intense heat. No such effect is to be expected in residential painting except under exaggerated conditions.

With the advent of many different types of paint, or protective and decorative coating, the problem was altered in some respects. Cellulose nitrate lacquers are violently flammable, while cellulose acetate lacquers are much less so. Varnishes tend to burn somewhat better than oil paints, but show no violent reactions that make them especially hazardous. The newer "rubber-base" paints, most of which are emulsions of polyvinyl acetate or acrylics in water, now much used for interior and some exterior decoration, are basically plastics and must be considered in this light. They offer no unusual dangers from fire, but are combustible under suitable conditions.

METALS

Metals, as fuels, are so rarely of importance that they may well be overlooked. Actually, most metals can be burned and many metals, when finely divided, are susceptible to direct combustion in air. Others can burn when they are in relatively massive form, but generally only in a very hot environment. Some metals are pyrophoric, that is, they burn spontaneously in air when finely divided and heated sufficiently. A metal like uranium, for example, is extremely hazardous when finely divided, because of its tendency to oxidize in air and the heat of the slow spontaneous oxidation may be sufficient to ignite a larger quantity of stored uranium powder. Iron powder is to a considerable extent pyrophoric, although without the general hazard that characterizes some other metals. The important qualification of the dangers associated with metal combustion is the state of subdivision. The finer the metal powder, the more likely it is to ignite, but rarely without supplemental ignition.

Perhaps the most important metals that sometimes serve as fuels in uncontrolled fires are *magnesium* and *aluminum*. Magnesium is not readily ignited unless in a rather fine state of subdivision. In a large fire, massive magnesium castings are capable of burning very fiercely with creation of an enormously hot fire. This particular tendency of magnesium, combined with its increasing use in industry, should alert the fire investigator to this possibility. In a case involving the burning of magnesium, it is more likely that the magnesium was ignited by a strong surrounding fire rather than that it initiated the fire. This will always be true when the metal is massive in form.

A matter that is of special concern with magnesium especially, and some other metals that are rarely encountered, is the fact that when water is sprayed on hot magnesium, large quantities of hydrogen gas will be evolved and will either explode or increase the fire by mammoth proportions. In fighting a magnesium fire, it is of the greatest concern that firefighters use foams, carbon dioxide, or other essentially non-aqueous materials.

Aluminum is much more difficult to ignite than is magnesium, because of its tendency to form a fine adherent film of oxide on the surface which protects the metal that underlies the film. Thus, in structural fires, it is very common to find much aluminum which has melted but not burned. Residence fires are hot enough to exceed the melting point of aluminum ($1220°$ F.) in a majority of instances but will rarely lead to actual ignition of the metal. Thin aluminum roofing may lead to this type of problem, because its thinness makes it more readily ignitable. Even with this material, it is more likely just to melt or to blow away in the updraft from the fire than to burn. If aluminum does ignite, it leads to an intensely hot fire because of its high heat of combustion.

The danger associated with burning of metals is confined almost exclusively to industrial plants, especially those in which the metals are handled in a finely divided state. Many disastrous fires of this type have been fed primarily by metal fuel. In general, such fires are so intensely hot and violent that they reduce the structure far more completely than is ever seen in a fire which is primarily fed by wood as fuel. Chemical laboratories and plants may contain enough of other alkali or alkaline earth metals to cause fire hazards.

FLAME COLOR

The color of the flame, especially in the very early stages of a fire, can be highly significant but will rarely be remembered or recorded with accuracy. It is clear that in most fires, the flame color is not likely to have great significance, because most structural, as well as outdoor fires, involve only vegetable material such as wood and some other organic construction materials such as asphalt, paint, and similar supplemental substances. For all these materials, the flame will be more or less yellow or orange in color, and only the amount and color of smoke that they give off will be related closely to the character of the fuel.

However, other colors of flame are sometimes seen in routine fires. When this is the case, it would be poor judgment not to give them close consideration. An illustration may be cited, in which an alcohol is used to kindle the fire. The lower alcohols burn with a relatively small blue flame. If such a flame were observed immediately after the fire started, it could well be indicative of the use of this type of material as an accelerant. It would also rule out hydrocarbon accelerants, all of which burn with a yellow, smoky flame. Natural gas, like the liquid hydrocarbons, burns generally with a yellowish flame, tending to be blue in the periphery. If adequate air is mixed with it, as is done in a heating burner, the flame is colorless to blue. As the weight progression of the hydrocarbons increases, the gases or vapors properly admixed with air burn with increasingly blue flames. Without the proper air mixture, they become yellow and smoky.

One gas that is of special interest at times is carbon monoxide. This gas is extremely flammable, although most of it is produced by air-poor flames, in which the combustion is not complete and the gas escapes before it is consumed by further oxidation. Carbon monoxide also burns with a blue flame. When, in a limited area the amount of fuel is in great excess over the amount of air to combine with it, there will be strong generation of carbon monoxide. This may escape to a region richer in air where it may be reignited, and ring the primary fire with blue flames. Such a situation will provide insight into the availability of excess fuel with limited air in the initial fire area. It is a situation that can exist with a large excess of volatile flammable liquid in a localized region, so that fuel becomes available in excess of the air available to burn it. The phenomenon is not likely to be seen when only a wooden structure is burning.

A good illustration of this effect was noted in a large hotel fire, the heavy part of which was in a subbasement, and fed by hydrocarbon. Firemen noted blue flames coming from a sidewalk entrance to the basement, while black smoke was emerging elsewhere. The blue flames were at a significant distance from the subbasement, and there was as yet little fire in the intervening space. The reasons for the existence of the bluish flame were not immediately perceived, and the only final explanation that fitted the facts involved a secondary burning of carbon monoxide of which the basement had to be very full because of limited ventilation.

Some other special flame colors may occasionally be of significance. Colored flames have long been generated for pyrotechnical purposes by admixture of certain metallic salts which generate the metals' spectra at flame temperature. Traces of sodium, present in most fuel, give a yellow flame, difficult to distinguish from the yellow produced in flames by glowing carbon particles. Much of the yellow that is so common in flames is partially or wholly due to sodium traces, but not specifically so. The significance of sodium in this connection is therefore minimal. Strontium salts, on the other hand, give the bright-red color used in many flares and decorative flame and explosion effects. Copper halides give an intense green color to the flame, potassium salts give violet, and barium salts a yellowish green. Many other variations of color will be characteristic either of the elements or of certain of their compounds. Clearly, these unusual colors of flame will only occasionally be of significance in the investigation of fires. However, it is conceivable that, under special circumstances, they could be the most important of all the indications. Assume that an intense fire has burned in the interior of a motor vehicle. This might have been due to various types of fuel material, one of which could be road flares. If an appropriate color had been noted in the early part of the fire, it could assist in locating the early fuel as flares. Yellow flames, on the other hand, could indicate more conventional fuels such as hydrocarbons. The investigator who is aware of the potential value of flame color in the interpretation of fire will be alert for any color that is unusual, or not explainable on the most obvious hypothesis. It could lead to more accurate appraisal of the cause and sequence of many fires.

It will also be noted that flame color, when not complicated by special elements that produce unusual colors, is related to the flame temperature. This in turn, is generally related to the oxygen admixture with the fuel, so that the color of the flame may be indicative of burning conditions as well as of the actual temperature of the flame.

SMOKE PRODUCTION

Although the character of the smoke produced by a fire is variable, observation of the smoke is one of the most helpful means of obtaining an indication of the type of fuel. When complete combustion occurs, most materials produce little or no visible smoke. Under these ideal conditions, all carbon is burned to carbon dioxide, a colorless gas with virtually no odor. Incomplete combustion of carbon leads to carbon monoxide, also colorless and odorless, but having very important toxic properties. In most fires, both gases are produced in large quantities and admixed. More important in this consideration is the fact that incomplete combustion of carbonaceous materials also produces various highly carbonaceous compounds that are colored. In the flames themselves, a great deal of elemental carbon (soot) will also emerge.

Hydrogen, universally present in all organic materials, burns to water vapor, which is a major constituent of all combustion gases. It does not contribute to the smoke characteristics, although it may at times be recognized as steam. Other elements often present in organic materials, such as nitrogen, sulfur, and halogens, contribute gases to the combustion products. However, only under very special conditions do these alter the color of the smoke. They may contribute strongly to its odor when present in larger than ordinary amounts. Such effects will be encountered in some industrial fires, but rarely in residential ones.

The *color of smoke* is almost entirely determined by the character and type of fuel and the availability of oxygen for complete combustion. Many materials, including all hydrocarbons, will never burn completely in the presence of excess air, without premixing of the fuel with air. This is even true of natural gas. As the hydrocarbon molecules become larger, as with oils, more air is required, and the means of mixing it with the fuel become more complex. Thus, in any ordinary fire requiring investigation, it is invariably true that materials such as hydrocarbons tend to produce a large quantity of very black smoke. Therefore, the presence of much black smoke, especially in the early stages of a building fire, is highly indicative of the presence and burning of a highly carbonaceous material typical of many fire accelerants.

Partially oxidized organic materials burning freely in the presence of air very often produce little or no colored smoke. Alcohols, for example, are in this category. If used as an accelerant, they will not be detected by observing the smoke. Wood, and most organic building materials, also fall in this category. They produce either a white or light gray smoke, except when the supply of air is greatly restricted. Even under these circumstances, they will not produce a heavy black smoke, such as follows the burning of hydrocarbons. Asphalt-like materials, also common in building construction, and such materials as tar paper, some paints, sponge rubber upholstery, adhesives, sealing compounds and some floor coverings, will generally produce a black smoke. These are more likely to burn late in a fire, so that the production of heavy black smoke late in a fire is not ordinarily an indication of the use of accelerant. However, it will at times give important information as to the course, sequence and nature of the fire.

The relation between the supply of air and the color of the smoke must not be overlooked. A fire burning a restricted space without adequate ventilation will always form more smoke and darker smoke than when the fire is ventilated. However, this is not so pronounced as to nullify the effect of the type of fuel. With hydrocarbon fuel, it will merely cause an increase in the quantity of black, sooty smoke. The experienced investigator will add greatly to his information about a fire when he considers the color and quantity of smoke.

Smoke color, and especially colors other than the neutral white, gray, or black, may give indications of special types of fuel that are being combusted (10). Such colors as red-brown, yellow, and yellow-brown generally signify that such items as nitrated or sulfonated materials, or other special chemical compounds are present. Photographic film will produce such effects, as will nitrocellulose, a common lacquer constituent. A smoke that has such an unusual hue should lead to a special inquiry as to the presence in the fire of materials other than the conventional building materials.

References

(1) Hodges, D. J. *Colliery Guardian*, **207**, 678, 1963.
(2) Johnson, A. and Auth, G., Ed. *Fuels and Combustion Handbook* (1st. Ed.). McGraw Hill, New York, 1951.
(3) Browning, B. L., Ed. *The Chemistry of Wood*. Interscience Pub., John Wiley, New York, 1963.
(4) McNaughton, G. C. *Ignition and Charring Temperatures of Wood* U.S. Dept. of Agriculture, Forest Service, Nov. 1944.
(5) Graf, S. H. *Ignition Temperatures of Various Papers, Woods and Fabrics*. Eng. Exp. Station, Oregon State College, Bull. 26, Mar. 1949.
(6) Angell, H. W. *Ignition Temperature of Fireproofed Wood, Untreated Sound Wood and Untreated Decayed Wood*. Forest Products Research Soc., 1949.
(7) Couzen, E. G. "Methods of Identification and Testing of Synthetic Resins and Other Raw Materials." In: *Synthetic Resins and Allied Plastics* (3rd Ed.), H. M. Langton, Ed., Chap. XVIII. Oxford Univ. Press, 1951.
(8) *Modern Plastics Encyclopedia for 1964* (vol. 41). Hildreth Press, Ind., Bristol, Conn., Sept. 1963.
(9) Lever, A. E. and Rhys, J. *The Properties and Testing of Plastics Materials*. Chemical Publ. Co., New York, 1958.
(10) Battle, B. P. and Weston, P. B. *Arson: Handbook of Detection and Investigation*. Greenberg, New York, 1954.

Supplemental References

Barr, W. M. *Combustion of Coal*. Norman W. Henley, New York, 1913.
Berger, L. B., Schrenk, H. H., Gale, J. A., Stewart, R. W., and Sieffert, L. E. "Toxicity and Flame Resistance of Thermosetting Plastics." *U.S. Bur. of Mines, Rept. Invest.*, No. 4134, 1947.
Simonds, H. R. *A Concise Guide to Plastics* (2nd Ed.). Reinhold, New York, 1963.
Simonds, H. R. and Ellis, C. *Handbook of Plastics* (2nd Ed). D. Van Nostrand, New York, 1949.
Troxell, G. E. "Fire Tests of Plastic Vents and Drainpipes." *Fire J.*, **60**, 52, 1966.

6

The Role of Pyrolysis

As discussed earlier, the greatest part of all flaming fires is the flame itself in which the combustion reaction is taking place solely between gases. This remains true even though the fuel that feeds the flames is solid, e.g., wood, cloth, paper, plastic, or even coal. How then does the solid fuel maintain a gaseous reaction in the flames that burn around it? The answer lies primarily in the phenomenon of pyrolysis, or heat decomposition reactions, which occurs within the solid fuel as a result of being strongly heated by the surrounding flames.

The phenomenon of pyrolysis has been known for a very long time, but the mechanism is only now being worked out for a considerable number of materials. However, with the exception of a number of pure compounds, it is still known to a limited degree only. To understand what occurs during the pyrolysis of fuel, it is necessary to consider that all practical fuels are organic in nature, and generally they consist of very complex mixtures.

Organic compounds are unique in nature because atoms of carbon, the element on which they are based, are capable of coupling with each other to form large and complex molecules. This is true not only for fuels, but for all materials that live, both animal and vegetable. Nearly all fuels of practical importance are vegetable in origin or are derived by decomposition, bacterial action, or geologic processes from originally living material, both animal and vegetable. This is presumed to be true of oil and coal as well as of natural gas. Wood, the most common fuel in ordinary fires, is the direct result of life processes in which very complex organic compounds are synthesized by natural processes. To date, nobody can write a structural formula for wood, although it is well known that its major constituent is cellulose, a very large molecular material synthesized from many glucose (grape sugar) molecules, in long chains of uncertain molecular size. In addition to cellulose, there are many other materials in wood; lignin is the most important quantitatively (up to about one-third) and there are various resins, pitches, and many other materials as well in limited and variable quantities. Certain soft woods, such as some of the pines, have large quantities of resins (the source of commercial rosin), while hard woods in general contain little or none of this type of material. These, on the other hand, tend to contain such materials as gallotannic acid, not obtained from the soft woods.

It may be noted that reasonably pure cellulose is rare, occurring in nature only in a few products such as the cotton fiber, which is about as pure cellulose as is obtainable in nature, although not chemically pure. Thus, the undyed cotton dress is reasonably pure cellulose. The structural wooden beam is composed largely of cellulose, but admixed with a large number of additional complex organic material that has undergone reduction to substances considerably higher in carbon than the original material that was reduced. It is, therefore, of a higher caloric value but less subject to pyrolytic decomposition.

Organic compounds (including the constituents of wood), when heated, are subject to extensive and complex types of degradation to simpler compounds which are more volatile and therefore more flammable than the compounds that were decomposed. The heating need not occur only in a fire. It may be equally accomplished by subjecting the sample to electrical heating in an inert gas, so that the decomposition products may be separated, identified, and studied. Such studies have assumed great importance in recent years in the elucidation of the mechanism of heat decomposition. For the purposes of fire investigation, it may not yet be important to understand the exact mechanism of the heat decomposition nor the products that result. However, it is important to understand the general facts of heat decomposition, because these are at the heart of comprehension of the basic nature of fire fed by solid fuel.

Wood, coal, paint, plastic, and many other solid organic compounds *do not burn*. When they are heated, they decompose to smaller molecules with greater volatility and flammability and to carbon. This is the process of pyrolysis, and it is the fundamental explanation of nearly all fires. It explains in toto the fact that it is fuel above the flames that is ignited and enhances the fire. Simultaneously, it explains why fuel below the flames does not burn readily—it is not heated enough to undergo pyrolytic decomposition into volatile gases that feed the flames. The gaseous materials formed by the pyrolysis of the large and nonvolatile molecules of the solid fuel are the materials that burn in the flame. Without them, there could be no flame, and often no fire would result. It also explains why grass fires may become crown fires in the burning of a forest fire. When a bush or tree is heated to the point that it exudes a large quantity of volatile gases that are flammable, the vapor cloud can burst into a sort of explosion, and the entire bush or tree is enveloped almost instantaneously in fire. This is the phenomenon that gives rise to the saying that a tree or bush "exploded" into fire. The material composing it simply was heated to the point that it decomposed by pyrolysis to form a large quantity of volatile, flammable gases, which when mixed with air underwent rapid combustion, similar to a small explosion. It also explains the mechanism by which fires are set at a distance by radiant heat from a large fire nearby. The radiation is absorbed until the temperature produced starts pyrolytic action, and temperatures rise until the combustion temperature of the pyrolysis products is reached. At this time, fire erupts all over the fuel exposed to the radiated heat.

An additional curious fact that is explainable only on the basis of pyrolysis products is the "fireball." It may be generated in a very intense fire and can travel through the air for appreciable distances while burning. The ball is formed by emission of quantities of pyrolysis products in excess of the air locally available to burn them. This means they accumulate in a limited region, burning primarily on the periphery. The strong updraft created by the fire carries them upward into the air where they continue to burn independently of the original fuel from which they were formed. They can drop if they move out of the direct updraft, and carry a mass of burning gases down into areas not yet on fire. It is not unusual for such fireballs to entrap personnel fighting the fire. Fatalities from this source are well known. Anyone who has viewed a very intense fire will have observed this phenomenon, often in miniature, as masses of flame that detach from the fire and rise into the air. In one large forest fire, several firefighters were killed when trapped in a pocket by a large fireball of this type.

LIQUID FUELS

A special case of the role of fuel in pyrolytic action arises from liquid fuels. Most of these materials, such as gasoline, kerosene, alcohol, and turpentine, are definitely volatile; when heated, they rapidly vaporize and the vapors burn. These materials are also subject to pyrolysis. The temperatures necessary to produce pyrolysis are generally higher than those which cause boiling, and the vapor becomes available to the fire without having to be pyrolyzed. It should also be remembered that liquid fuel of this type cannot be heated significantly above its boiling temperature, because it will all boil into the fire. The loss of heat of vaporization will keep it cooled to the boiling point until all the liquid is gone.

Under appropriate circumstances it is possible also to pyrolyze volatile liquids. However, the arrangements for doing so do not exist in the fire that requires investigation. Some liquids which are not very volatile are subject to pyrolysis in a manner not greatly different from solids. This is true of asphalts, which are true liquids even though generally encountered in a solidified form. Various liquids that are rarely encountered outside of the chemical laboratory fall into this category. For all practical purposes, it may be considered that the liquids normally of interest to the fire investigator are not pyrolyzed in the fire but distilled into it. Thus, it is only solid fuels, especially wood, that are of immediate and direct interest.

WOOD PYROLYSIS

Relatively little detailed study has been made of the pyrolysis of wood. Perhaps the most useful treatment of the subject is that of Browning (1). The applicable statements follow:

"The thermal decomposition of wood, when carried out in the absence of oxygen, results in the production of combustible and noncombustible gases and vapors, and a residue of charcoal remains... Thermal decomposition also occurs during the combustion of wood and establishes the nature of the combustion process.

"Wood is stable toward heat, except for loss of hygroscopic water, up to about 100°C. (212°F.). As the temperature is further increased, carbon dioxide, carbon monoxide, hydrogen, and water are formed by the chemical decomposition of the wood constituents. Between 100 and 250°C. (482°F.), decomposition causes the wood to darken in color and lose its strength, although the structure is retained. At higher temperatures, up to 500°C. (932°F.), carbonization occurs and additional volatile materials are lost. The reaction (in the absence of air) becomes exothermic at about 275 to 300°C. (527 to 572°F.). The decomposition reactions of lignin and cellulose become exothermic at about 270 and 300°C. (518 and 572°F.) respectively. Pyrolysis of alpha-cellulose at about 300°C. (572°F.) and of lignin at about 400°C. (752°F.) has been reported. Hemicelluloses decompose at a considerably lower temperature.

"The pyrolysis is largely completed at about 500°C. (932°F.), and the residue of wood charcoal remains. However, charcoal produced at this temperature still gives off a considerable quantity of noncondensable gases on further heating at 800°C. (1472°F.).

"In the range of wood distillation or carbonization the products formed can be classified broadly as noncondensable gases, pyroligneous liquor, insoluble tar, and charcoal. The products obtained by laboratory pyrolysis over the range of 250 to 350°C. (482 to 662°F.) were approximately 27.5 percent water, 10 percent noncondensable gases, 2 percent acids and methanol, 5 percent tar, and 8 percent settled tar. Over the range of 350 to 450°C. (662 to 842°F.), the products were 4 percent water, 3 percent noncondensable gases, 2 percent settled tar, and 0.5 percent acids and methanol. The noncondensable gases consisted largely of carbon dioxide and carbon monoxide, with smaller quantities of hydrogen and hydrocarbons."

It will have been noted that pyrolysis, as studied experimentally, is done normally in the absence of oxygen, which is not true completely with an open fire. Thus, in the fire, there are both combustion and pyrolysis occurring. The combustion is primarily in the flames, and the pyrolysis is in the solid fuel. Actually, on the surface of the solid fuel, there is a considerable deprivation of oxygen since it is being used by the flames before it can reach the solid surface. In addition, the pyrolysis is proceeding in layers below the surface of the fuel, where there is nearly total deprivation of oxygen. Thus, the basic facts of pyrolysis are entirely applicable to the solid fuel, despite some interference from actual combustion.

RELATION OF ORGANIC STRUCTURE TO PYROLYSIS PRODUCTS

Organic chemists have made considerable progress in tracing the relation of the chemical structure of the heated solid to the products that emerge when it is heated to the point of pyrolytic decomposition. In fact, this process is sufficiently reproducible that analysis of the pyrolysis products, e.g., by gas chromatography, now allows the chemist to predict many features of the material pyrolyzed when the structure of that material would not otherwise be known. The mechanism of pyrolysis involves the breaking of chemical bonds, the weakest being broken preferentially, with formation of free radicals, that is, pieces of molecules with unsaturated or uncoupled electron pairs. These free radicals may: (1) disrupt further to form unsaturated compounds, elemental gases such as hydrogen, or smaller radicals; (2) combine with each other to form new compounds not originally present; or (3) react with the original substance. Because of the tendency of free radicals to recombine into new and more stable configurations, the compounds formed may not appear to have any relation to the parent material. For example, benzene is formed in the pyrolysis of most organic material, even that which contains no benzene rings initially. Other simple compounds commonly formed include such gases as methane, ethane, toluene, and other simple flammable gases and liquids.

Because of the relation of original chemical composition and structure of the fuel to the products produced by pyrolysis, it is suggested that study of the materials that often condense some distance from the actual fire would allow a reasonable determination of the general nature of the material burned. For example, in a structural fire, it is not uncommon for several rooms to be unburned but "heat damaged." Here, the general effect is discoloration of walls, tiling, and mirrors, or window glass with much soot, streaks of condensed moisture, and a dark, greasy film. Much of this material is the relatively less volatile pyrolysis products that have distilled out of the fire without burning. Apparently, little or no investigation has been made to determine whether or not the analysis of such deposits could give information as to the nature of the fuel that was burning. Such a determination could be important in distinguishing between the soot formed from an accelerant such as gasoline and the similar soot formed by burning of asphaltic roofing, sealing compounds, paint, and other similar material. This determination could be most helpful sometimes in distinguishing between arson, in which accelerants are used, and an accidental fire, in which the fuel is of a cellulosic nature that produces quite different pyrolysis products.

EFFECT ON IGNITION TEMPERATURES

The ignition temperatures quoted in the literature are largely for liquid fuels, in which pyrolysis is not expected to occur, because of their relative volatility. Determination of an ignition temperature of solid fuel is difficult at best and with only limited meaning. Assume that a piece of wood is heated until it finally bursts into flame. When this occurs, it is not the wood that is burning but the products of pyrolysis of the wood, and the ignition temperature is related to that of the gaseous products, as well as to the pyrolysis temperature of the wood which is not a definite value. It is, to a large degree, a progressive effect as the temperature rises.

Another complication arises in that the products of pyrolysis are multiple in nature; each such product has its own ignition point and its own range of composition within which it will burn. Thus, one product will reach a suitable concentration ahead of another that is formed to a lesser degree or has a higher minimum percentage in air that can burn. In such an event, only one of the pyrolysis products controls the apparent ignition point, even though another less plentiful product might have a lower value. If the heating is done rapidly enough that several products reach their limits of flammability about the same time, the lowest ignition point of a compound in the group is the one that will determine the observed ignition temperature.

It is common experience that woods of different species vary considerably in the ease with which they are ignited. Coal is more difficult to ignite than wood. Here again, the explanation resides largely in the ease with which the various fuels can pyrolyze. Some materials start to decompose under heating earlier than others. Coal has so large a carbon content with so little chemical structure, that it is far more difficult to pyrolyze than any wood. The factor which is chiefly responsible for the differences in ease of ignition is the ease with which pyrolytic decompositions can take place. Hard coal and especially coke or charcoal are nearly all carbon, and pyrolysis is a very small part of the mechanism by which they are ignited. These materials have to be heated sufficiently to initiate a glowing condition in which surface combustion starts before they will burn spontaneously. Whatever flame results is largely that of carbon monoxide, although some compounds subject to pyrolysis do exist in minor amounts in these fuels.

PAINTS AND OTHER SYNTHETICS

The role of pyrolysis in the ignition of paints, plastics, and similar complex organic materials is of the greatest significance. Materials such as these are ordinarily quite susceptible to heat decomposition with formation of more or less gaseous products that burn readily. It would be a mistake to assume that because this is true, painted wood is necessarily more susceptible to combustion than unpainted wood, or that plastics are more of a fire risk than wood. Paint may be quite fire retardant, both because of lesser ease of pyrolysis and especially because of inclusion of mineral (incombustible) materials, that shield the paint from radiant effects while remaining stable themselves. Many plastics also have the property of melting at rather low temperatures, and the melted plastic may flow away from a localized source of heat without giving enough pyrolysis products actually to contribute significantly to a fire. Here, the evaluation is relative, because plastics are for the most part capable of supporting very intense fire when heated sufficiently. The fluorocarbons are an exception, as is discussed elsewhere, and wide variations exist between different types of plastics other than the fluorocarbons. The important point is that the role of pyrolysis must not be neglected in any consideration of the flammability of a particular fuel or mixture of fuels that are subjected to combustion phenomena.

PYROLYSIS WITHOUT COMBUSTION

When compounds and mixtures are subjected to the temperatures at which pyrolysis takes place, they not only break down with formation of gaseous (and generally flammable) products, but it is normal for a highly carbonaceous residue to remain after the pyrolysis. Charcoal is such a residue from wood, coke from coal. In structural fires, the effect of pyrolysis without combustion is often noted. Perhaps the best illustration is wall paper on a plaster wall. When there is a large fire in the room, a black coating remains on the plaster, often with the pattern of the wall paper (mineral) still visible. It will be considered that the paper has been "charred," but the real effect is that the paper was pyrolyzed, the gaseous products burned or lost in the atmosphere, and a carbonaceous residue left behind.

Whether it be wall paper or charred wood, the effect is the same. Both paper and wood were subjected to temperatures sufficient to produce pyrolysis; the products burned, and the carbonaceous residue was left behind to screen and protect the underlying surface from further effect of the radiant (and other) heat. The effect of this layer of char is primarily insulation, both against conducted and radiated heat. The investigator will often see boards and other timbers burned only on one side with the other appearing fresh and with no trace of char. Here, the insulation produced by the char layer has protected the remaining wood from the effect of the heat. Only when the fire penetrates so as to attack the board or timber from both sides is there a high probability that the timber will burn through. In such an instance, it must be considered that flames are very hot and produce quite rapid pyrolysis, with all of its consequences.

It is clear that a knowledge of pyrolytic mechanisms is basic to the understanding of fires, because pyrolysis is one of the most fundamental accompaniments of all fire and a contributing cause of many. It is especially significant in its role in spreading fire, and it is a major contributor to ignition of fires as well. All that is necessary is sufficient heat to break down the chemical structure of the fuel with formation of more flammable materials in the gaseous state. Because these generally have a lower ignition temperature, as soon as they reach the flammability limits at a temperature above the ignition point, flames must result and a fire is underway.

Reference

(1) Browning, B. L., Ed. *The Chemistry of Wood*, Interscience Pub., John Wiley, New York, 1963.

Supplemental References

Eickner, H. W. "Basic Research on the Pyrolysis and Combustion of Wood" *Forest Prod. J.*, **12**, 194, 1962.
Tang, W. K. and Neill, W. K. *J. Polymer Sci.*, **Part C**, (6), 65, 1963.

7

Fire Patterns of Structural Fires

With few exceptions, fires, however large, start with a small flame, such as a match, candle, or lighter, or a spark before the flame. It is the point and source of ignition that are the keys to be determined by the investigator, regardless of any holocaust that may have followed this tiny flame. These determine the origin of a fire and must be primary in any proper investigation. An investigator who spends a large proportion of his time surveying what are obviously the later stages of the fire is wasting much of that time, because such survey cannot answer the question he is investigating. The insurance adjustor may be interested in the final stages of the fire in which most of the damage occurred.

72   *Fire Investigation*

This is appropriate to his task. However, when there is a need to determine why and how the fire occurred, the investigation must be concentrated on the tiny initial flame.

Since it is of the greatest importance to trace the behavior of this small flame as it grows into the large fire, the following simple rules of fire behavior are listed.

1. Hot gases (including flames) are much lighter than the surrounding air and, therefore, rise. To force them to burn any other way requires wind currents of some magnitude. Thus, in the absence of such an abnormal situation, the fire will always burn upward.
2. Combustible materials in the path of the rising flame will be ignited and increase the fire, causing greater flame volume and more vigorous rise of the flame.
3. In most instances a fire can only develop when fuel is above the initial flame so as to increase its volume. Otherwise the fire will quickly burn itself out.
4. Variation of this upward direction will occur with application of lateral air currents that will deflect the flame, or in the presence of strong drafts that can force it downward. Fuel in the path of the flame is still necessary, even in this unusual situation.
5. Lateral spread of fire will occur to a limited extent in an open environment, but can occur rapidly when an obstruction to upward burning is encountered. Thus, ceiling fires always tend to predominate in a building, because the rising fire striking the ceiling can no longer rise and the trapped hot gases and flames rapidly attack whatever fuel the ceiling affords.
6. Fire always seeks chimney-like configurations, because here the rise of the fire is enhanced. Stairways, elevators, dumb waiters, interiors of walls, and similar vertical openings carry flames generated elsewhere and, when ignited, always tend to burn more vigorously than other areas.
7. Downward, as well as lateral spread, is on occasion markedly increased by the presence of highly flammable coatings or bonding material, or by the presence of extraneous materials utilized in treating wood surfaces, as in painting. When this type of situation prevails, the fire pattern may be greatly modified, and the special condition must be recognized so that misinterpretation does not result. Not only is the occurrence of such a situation relatively rare, but its recognition is not difficult, either from the pattern itself, or from extraneous information, and often both.

IMPLICATIONS

Every fire forms a pattern that is determined chiefly by the configuration of the environment and the availability of combustible material. Because of the upward tendency of every fire, some type of inverted conical shape is characteristic, the apex at the bottom being the point of ignition, with the fire rising and spreading. Naturally, this pattern will be altered by the presence of obstructions, or of readily burned fuel in localized areas. Thus, interior fires often appear to have very complex patterns, the result of configurational complexities rather than of any unusual property of the fire.

Some illustrations of the type of complications in patterns that may be encountered are in order. Many walls will burn through slowly or not at all. Partitions or small sections of wall surface may be constructed of thin material which burns through and starts general distribution of the fire on the other side of the wall. Holes in walls, either placed intentionally or produced accidentally, may set up unusual draughts which allow an uncommon lateral spread. Even a mouse or rat hole through the base of a wall has been known to produce this type of effect. Another very common complication arises from presence of stairways, elevator shafts and dumb waiters, all of which tend to burn very fiercely because of the excellent ventilation. The intense burns noted in such areas may well distract the investigator from following the fire pattern back to its point of origin. It may be at a considerable distance in space and relatively much less intensely burned. Where there are numerous horizontal structures, such as ceilings, whether they are combustible or not, rapid lateral spreading may occur, so that deviation from the characteristic and uncomplicated cone is produced. This is very common in structural fires.

The important item in every fire pattern is the apex of the cone at the bottom. Here is where the fire started as a small flame, and here is where the source of the ignition may sometimes be found. It is this point that is the goal of the fire investigator. However, complications are sometimes encountered in locating this point.

DEVIATIONS FROM NORMAL PATTERN

In discussing the simple fire above, it was assumed that the fire started at a point. This is not always a correct assumption. If a liquid accelerant is spread on a floor, the entire area so covered catches fire at about the same instant, and the point is now an area. As will be developed in a later section, this complication need not seriously interfere with locating the origin of the fire.

Another more serious deviation from the normal pattern is often encountered in large fires where secondary ignition from falling material may occur. Collapsing structures may drop so much flaming material on the point of origin as to obliterate it or even to start secondary fires at levels lower than the initial point of origin. This latter event is not very common, since the level on which the fire started is generally the lowest level that is reached by falling material. As a rule, nothing below this level has been destroyed and it therefore stops any falling materials. The problem of obliteration of the point of origin can be much more serious. But even this possibility is rarely a point of difficulty, because when a structure collapses, it generally smothers the fire on which it falls and preserves the point of origin excellently. Only digging is then required to uncover the low burn which corresponds with the point at which the fire started.

LOW BURNS

Any low point in a burn pattern should be investigated as a possible origin. However, when such a low point is found, it usually shows that an intense fire existed there, rather than the minute flame discussed above. This effect results from a combination of factors which may be outlined as follows:

1. Any small source of ignition which failed to find the requisite conditions for development into a large fire is extinguished spontaneously. This type of event is not uncommon. A cigarette or a match thrown into a waste basket will often cause minor charring of papers contained therein, without initiating a larger fire. In fact, many small fires are self-extinguishing.
2. When the small source of fire finds a favorable environment, it rapidly grows into a larger fire that can become a general fire.
3. Because the point of origin is likely to burn longer than the fire that develops from it, more time is allowed for it to produce an impressive degree of burning.
4. When the fire is aided by the presence of accelerants or kindling materials of any kind, an unusually intense fire may result at the point of origin, through design as well as chance.

The lowest point of burn must always be inspected with the greatest care. This is especially true if one is to determine the cause, as well as whether or not the fire was of incendiary nature, deliberately set, or accidental.

Configuration at a Low Burn

In many instances, the point of lowest burn is a floor surface or a region directly under the floor. These points are sometimes difficult to evaluate and often lead to error in interpretation. For example, there is a hole burned in the floor in a region away from all walls or other objects which could carry the fire upward by providing fuel in the path of the flames. It is not uncommon for the investigator to assign the cause to the use of a flammable liquid. Such an interpretation is more often incorrect than otherwise. On a tight floor, it is always incorrect, unless holes or deep cracks are present. Lacking such conditions, *flammable liquids never carry fire downward*. The floor surface which holds them cannot be heated above the boiling point of the liquid applied, as long as liquid is present. This temperature is far too cold to ignite the floor. However, if the liquid itself can penetrate to a lower level by way of cracks or holes, the fire can then be generated below the surface and will burn holes in the floor. This point is of the greatest significance.

There may be some or no observable floor burns, but the wall, or some other vertical and flammable surface, is burned to the level of the floor. This situation is generally consistent with the use of liquid flammables, applied at the bottoms of such vertical surfaces. When the burn is even with the floor surface, it is highly probable that a liquid was used, since other sources of intense local combustion, such as trash, rarely if ever burn right to floor level. The flames from a pile of flammable trash burn upward in the typical inverted cone and contact the wall higher than floor level. Liquid, on the other hand, can soak into the crack that separates the wall from the floor and take the fire to the very bottom of the wall.

Discoloration and minor charring of a floor surface can result from certain situations concerned with liquid flammables. The liquid itself, through its solvent action on waxes, etc., of the floor, may slightly discolor the surface. At the edge of a burning pool of liquid, the flame has limited exposure to the dry floor surface. With persistent liquids, this can produce a local but superficial scorch. Gasoline, turpentine, and similar solvents sometimes produce this effect, while alcohols, acetone, and similar, more volatile liquids do not. In no instance is this effect to be confused with the actual burning of a hole in the floor.

The shape and configuration of floor markings such as these are often used as arguments in favor of the application of flammable liquids to start the fire. A liquid thrown on a floor tends to assume a rounded or oval shape, distinguishable from that of a trash fire. When it applies only to discoloration or very light charring, the interpretation may be valid. However, there are many causes for rounded configurations that have nothing to do with liquid flammables, and care must be taken in reaching such a conclusion.

Holes Burned in Floor

When a hole is present in a floor, and no other point of the fire is lower than this, the investigator's attention is drawn to the hole. In the interpretation of holes of this type, it is most important that the investigator pay special attention to all of the possibilities. As developed above, holes are burned in floors by liquid flammables only when the liquid itself can penetrate below the floor surface. However, there are several other causal factors that can readily burn a hole in a floor. Some simple experiments on the part of any doubtful investigator should always be made before stating any final conclusion about holes burned in a floor. For example, flaming draperies burn off their supporting rods and fall, playing flame directly on the floor. This situation is not infrequently found in residence fires, where the holes are uniformly below windows and often outline very well the region over which the draperies were hung. Extending this point, any material which is flaming vigorously and dropped on a floor may be expected to burn a hole in it.

Very hot objects which have a high heat capacity, such as a red hot chunk of metal, will burn down into a floor and may penetrate it. For example, molten aluminum from an aluminum roof may produce this effect.

The fact that flammable liquids cannot burn through a tight floor, while flaming solid materials can, is not a sufficient reason to assume invariably that such a hole must therefore have been caused by such flaming or very hot material. Even a tight floor may have isolated holes or cracks which extend through it sufficiently to allow penetration of flammable liquids. Such orifices will generally be totally or partially destroyed by the resulting fire. It is necessary at this point to illustrate the preceding principles.

1. A large hole, some two by four feet in general dimensions, was burned through a floor immediately in front of and partially under the front of a refrigerator standing in a concrete-floored recess, and partially overlapping the wooden floor. The wooden floor was covered with a tight covering, and was itself made of tongue and groove boards placed over a rough floor. It is virtually certain that the floor was tight to its edge. It was certain that the refrigerator was not involved since it was empty, disconnected, and damaged only by the flames from the burning floor. The tightness of the floor would eliminate penetration of liquid. However, the edge of the floor abutted the concrete, and a definite gap occurred here through which liquid could flow. No other source of the burn could be found, and the flames had burned from the ground upward, heavily involving the lower sides of the floor joists. In addition, there were flames underneath the floor in regions where the burn was incomplete. The use of liquid which flowed under the floor at the junction with the concrete was the only tenable conclusion.
2. A number of holes were burned in the living room floor during a residence fire. These were attributed to liquid flammable material by the fire department, who considered the fire to have been an arson attempt. The floor was a typical hardwood floor of good quality, completely tight. Analysis of the burns showed that every hole was in front of either a window or a door which had a glass pane section. Questioning elicited the fact that the lady of the house used heavy drapes over all glass openings, and the pattern of burn was in complete agreement with drapery fire.
3. A heavy fire in a dock warehouse containing railroad tracks and supported by very heavy timbers and piling had a floor hole some ten feet in diameter. The fire had burned through timbers with dimensions in excess of one foot. In this instance, the rough planking of the floor allowed the presence of wide cracks, and the violence of the fire that burned upwards below this

floor was consistent only with an extended fire from below. This could have come, in this instance, only from liquid fuel poured through this floor. The availability of drums of liquid flammables, and the absence of any source of falling material, lent credence to the idea that this conflagration was a set fire. Further, the fire was from below rather than from the floor surface down.

The *primary rules* that emerge from this consideration and the illustrations are as follows: (1) a hole burned upward through a floor shows the presence of flammable material below the floor, often liquid in nature; and (2) when the hole is burned downward, it results from flaming material dropped from above the floor, and it will show no charring of consequence past the edge of the burn on the bottom of the floor surfaces. The fire from below chars and consumes the exposed bottom of the floor over a definite distance from the margins of the hole. Examination of the lower surface of the floor and its supporting structure past the margin of the penetration will show whether the fire burned upward or downward. This is the critical determination that must be made.

COATINGS AND SPECIAL TREATMENTS

As will be discussed below, paints tend to retard fire because of their high mineral pigment content. Varnishes lack pigment ordinarily and they contribute to the fire, but usually they do not alter greatly its interpretation. Nitrocellulose lacquers, being highly flammable, are expected to increase rapidly lateral and downward spread and to involve all surfaces so covered. Likewise, the adhesive used in making some plywoods is exceptionally combustible or is affected by the heat so as to open the laminations and accelerate the spread of flames. Washing of walls with solvents will sometimes allow enough penetration of the solvent into the wood to increase the flammability dangerously. All of these effects will tend to result in charring or total burning of all such surfaces, regardless of the additional pattern of the fire. When such total involvement of the surface is encountered, the rule of low burns will not apply to these surfaces.

ROOF AND ATTIC FIRES

In residences, especially, it is not uncommon for a roof to burn away or for a combined attic-roof fire to occur. For this reason, attic and roof fires are often confused, and this can lead to misinterpretation. Actually, a *roof-started* fire is totally different in behavior than an *attic-started* fire. In both instances, the absence of fire on a lower level will define the fire as being one or the other, but often it is difficult to determine which.

An attic fire is started *under* the roof and will inevitably involve all of the roof that is accessible to the spread of flames beneath it. Thus, it is likely that the entire roof is either burned away or that destruction is very widespread. If there are regions of the roof which remain intact, an attic-started fire was burning under them to some extent. Widespread burning in the attic is always indicative of this type of fire.

A strict roof fire, on the other hand, must have been ignited from the exterior, otherwise it would develop as an attic fire. Sparks from chimneys, or similar causes, may ignite shingles or shakes on a roof, resulting in an *exterior* fire. This will follow the usual inverted cone pattern on an inclined roof and ultimately will penetrate the roof with formation of a hole into the attic below. It is at this point that the main distinction between an attic-started and a roof-started fire becomes apparent. The first hole that burns through the roof ventilates the fire, drawing air upward through the hole. This generally confines the fire to the exterior and to parts of the roof immediately adjacent to the hole. Here, *air is being drawn from the attic to the fire*, which in most instances suffices to prevent a secondary spread as an attic fire. The major burn is on the outside of the roof, not under it.

In an attic fire, there is almost always a great deal of burning under all parts of the roof before the latter is finally breached. Since ventilation is limited, spread of the fire is inevitable. After a hole is chopped in the roof, to ventilate the fire under it, that fire can then be controlled much more easily because the flames are funnelled toward the hole which ventilates them.

Although the destruction may not be significantly different in the two cases, the investigation is greatly affected. This is because the roof fire is caused by *external* ignition, and the attic fire is ignited *within* the structure.

INTERIOR FIRES FROM EXTERIOR SOURCES

Fires within structures need not invariably have their origin in the interior. Buildings are frequently consumed as a result of a fire which is initially outside of the structure. Such sources of fire may be roughly classified as follows:

1. Fire from an adjacent building, most frequent in congested areas.
2. Grass or brush fires exterior to the building.
3. Trash fires which are out of control.
4. Arson by means of trash or accelerant ignition against the side of the structure, under porches, around doors, etc.

When the fire is initially outside the building, it may involve exterior surfaces and never penetrate to the interior. Such a fire pattern is readily evident, and the cause simple to determine. However, special circumstances may exist to complicate the situation. An open door or window, combined with wind and burning fragments from a nearby fire, may start the fire on the interior with no burning of the exterior structure. This fire would have an interior origin which would be traced in the conventional manner. The determination of cause in such an instance may have to rest on such circumstances as the existence of a nearby fire, the wind, the opening, and the absence of any other reasonable interior cause. While such instances are not common, they can be quite troublesome to the investigator.

It must also be considered that the most common type of transmission of fire from an exterior source is by burning material which settles on a flammable roof. The resulting fire is not initially an interior fire. So many variations in the mode of transmission of fire from point to point, depending on the ambient conditions, are encountered, that care and thoroughness on the part of the investigator are essential. All possible factors must be considered and each weighed comparatively and in light of the environmental conditions that can be established.

TRACING THE PATTERN

A systematic approach to the study of fire pattern is possible and should be followed. It may be outlined, generally, as follows:

1. Upper portions of the fire may be disregarded at this stage of investigation.
2. Low burns are systematically sought, making sure that the very lowest is located, and that all burns on approximately the same level are found.
3. Each low burn is analyzed as to the spread of fire away from it. This is accomplished by noting the direction of predominant fire, as shown by depth of burn. The directions involved must be noted, as well as the areas of burn that could have been initiated as a small fire at this low burn.
4. Having made a survey of the low burns and the fire that may have resulted from each, over-all gross pattern is studied to ascertain the way the fire burned after it developed as a large fire.
5. Extraneous factors, such as wind direction, presence of chimney action, and the like are now fitted into the pattern to determine how much of the large fire was influenced by such factors as these.
6. At this stage, the low burns are assessed in their relation to the total pattern. As a general rule, only one will be consistent with being an origin of the fire. If there are two or more, it is reasonably certain that a set fire occurred.
7. If only a single low burn is consistent with being the point of origin of the fire, causes for this low burn are sought, as outlined in other sections of this volume.

Depth of Charring

In studying the pattern of a structural fire, variations in the depth of the char will inevitably be noted. Some investigators consider that this feature of the fire is of primary importance; they make measurements with the idea of determining the length of time the fire burned at this point. It is evident that other factors being constant, there should be a direct relation between char depth and time of burning. However, to place more than casual emphasis on this point may well lead to failure to diagnose the situation fully.

Not only the time of burning, but the flame intensity, which varies locally to a great degree, will often be involved. Different species of wood will also char at somewhat different rates. The specific gravity of the wood is a major factor in this effect, as it also is with regard to the rate of spread of a fire. The adherency of the char is also an extremely important variable with certain species such as redwood. Even the dimensions may have an influence because of their variable effects on heat capacity and conductivity. To complicate the practical picture further, misinterpretation may follow from the fact that in fighting the fire the application of water will inhibit charring in one area while close by, there may not have been any water applied, although the general fire intensity was similar.

Few reliable controlled data are available as to the rate of penetration of fires to produce charring, although frequently statements of fire investigators on this point are encountered. As an illustration, the following statement was given in a fire report: "As a rule of thumb, free burning dimension lumber will char at the rate of one inch in 45 minutes." Bird (1) gives the penetration rate of hardwood as one-eighth inch in five minutes. Such figures may be of help, when the investigator takes into account the numerous variables, and especially the adherency of the char which serves to inhibit the penetration progressively with time.

Rate of Fire Spread

A more important factor, and one that has been studied to a much greater extent than depth of charring, is the rate of spread of fire along exposed wooden surfaces. It also is of limited utility to the investigator seeking the origin of a fire. However, it has numerous implications for the insurance investigator and for the builder who wishes to avoid unusual fire hazard. Some of the factors which influence the rate of spread are species, density or specific gravity, ring orientation, and moisture content. The Forest Products Laboratory of Madison, Wisconsin, has conducted numerous tests along these lines. Fons, Clements, and George (2), especially, have reported useful controlled data. As expected, the rate of spread was greatly influenced by arrangement of the wood sticks, and "rate of speed increased rapidly with decreasing moisture for specific gravity less than 0.45 and moisture content less than 10 percent." Other publications dealing with this effect include Steiner (3), Cross and Loffus (4), and Kollmann and Teichgraber (5).

References

(1) Bird, G. I. "Fire Resistance of Floors and Ceilings." *Joint Fire Research Organization Fire Note 1*, 1961.
(2) Fons, W. L., Clements, H. B., and George, P. M. *Scale Effects on Propagation Rate of Laboratory Crib Fires.* 9th Symposium (International) on Combustion, Cornell Univ., 1962, Academic Press, N.Y. and London, 1963.
(3) Steiner, A. J. "Burning Characteristics of Building Materials." *Fire Engineering*, 104(4), 264,
(4) Cross, D. and Loffus, J. J. "Flame Spread Properties of Building Finish Materials." *Bull. Amer. Soc. Testing Materials*, No. 230, 1958.
(5) Kollmann, F. and Teichgraber, R. "Testing the Burning Properties, Especially the Resistance to Combustion, of Boards of Wood and Wood-Based Materials." *Holz, Roh-u. Werkstoff*, 19, 173, 1961.

8

Fire Patterns of Outdoor Fires

Although occasional small, localized, outdoor fires may occur in the vicinity of human habitation, outdoor fires are chiefly found in forests, brush, and grasslands. Such fires feed on trash, lumber, or similar materials. Sawmill fires, which are relatively common, would also be classified as modified exterior fires rather than building or structure fires. In their fundamental character, exterior fires are not different from interior structure fires. They often are different in cause, and interpretation of their spread is modified greatly by the predominant horizontal spread in the open, as compared with a predominant vertical spread in structures. They differ also by various types of impediment and irregular distribution of fuel.

FOREST AND BRUSH FIRES

Trees and brush are so similar in their manner of burning that they may be considered as variations of the same category. Here there will be distinguished two fire characteristics:

1. *Ground fire*, feeding on low material such as grasses, weeds, low brush, and dry leaves but not burning very high. Fires such as these are not uncommon in virgin forests with high trees. Such fires differ little from grass fires, but they involve tree trunks, which are absent in the grass fire.
2. *Crown fires*, involving the tops of trees and high brush. Fires in this category are violent and often are termed "wild fires." They are difficult to fight because they tend to spread with great rapidity and produce excessive amounts of heat that makes approach to them difficult. The determination of the pattern of such a fire is apt to be more difficult than that of the ground fire.

EFFECT OF WIND AND TERRAIN

These influences are treated together, not because of any obvious relationship but rather because the effects become so intertwined that their separation is not possible. From the standpoint of investigation, where the fire burns in its later phases is relatively unimportant. The *critical determination is the point of origin.* When the fire started, it was not a large fire but a very tiny one, and if the location of that tiny fire is pinpointed, its progress from that point is of interest primarily in assessing the total damage. It is in this process that consideration of the wind and the terrain is of critical importance.

Although wind is to some extent independent of both terrain and the existence of the fire, it is also partially dependent on both factors. For example, there may be a prevailing wind coming from the west at the time of a fire. As the air currents intercept hills, valleys, and broken terrain, deflections occur; at one point the wind may be from the northwest, at another from the northeast, and at still another, even from the east. The latter effect is largely due to wind striking a rapidly ascending slope which deflects it upward. This draws air currents upward on the other side of the slope, thus creating backward wind direction at that locality. Thus, it is not sufficient to consider prevailing wind direction as determinative of the direction of movement of a fire at a particular time and place. In hilly and broken terrain, it is not uncommon to find that the fire has progressed in a number of directions, depending on the location studied within the burned area.

One of the most important sources of wind associated with fire is that which fire itself creates. To a large extent this effect is independent of prevailing wind direction. The heat from the fire causes the rapid rise of combustion gases and the heated air in the vicinity. This produces a partial vacuum which causes air to flow into the base of the fire—an invariable result of every fire. With a large fire such as may occur in the forest, this effect is great and a strong wind is created by the fire itself. The result is air flowing from every possible direction to the base of the fire. On level ground, such wind will enter the fire from all directions, thus having only local direction at any given point. On a steep hillside, the wind will flow toward the base of the hill and upward. Ambient wind will act only to modify this effect. It will require a strong wind to do more than deflect the fire's own air direction, but it can occur.

It is evident from the above considerations that fires tend strongly to burn uphill, and only unusually strong downhill wind will reverse the tendency. While such a reversal is rare, some of the most costly fires have been associated with this exact effect. Thus, every forest or grass fire must be studied first with respect to the *direction of burning*, which is determined almost entirely by the combination of terrain and wind directions, these having mutually interdependent relationships. The rules may be stated as follows:

1. Fire, uninfluenced by strong wind, will always burn uphill in a fan-shaped pattern.
2. On level ground, in the absence of wind, fire will spread from a center in all directions, but its spread will be inhibited by the wind it creates, blowing into the base of the fire from all directions. Such a fire will spread very slowly.
3. Ambient wind will modify the pattern by adding an additional spreading component, so that the fan-shaped pattern on the hillside will deflect to one direction or the other, and a predominant direction of travel will be created on level ground.
4. In the uncommon circumstance of a strong downhill wind, the fire will burn down the hill only to the degree that the ambient wind can overcome the fire's own tendency to burn uphill.
5. In a fire having an extended perimeter, the direction of burning may vary locally in almost any direction, depending on the interdependence of the terrain, the air currents created by the fire itself, and the ambient wind.

INVESTIGATION

An exterior fire is rather easily investigated, once the above general principles are understood. In many instances, there is initial information that is helpful in determining the general locality of the fire before it became very large. In all such instances, the investigation can be limited to the area in which the early part of the fire was burning, regardless of how far it may have spread later. The direction of burning at each point examined is determined by noting on which side of stems, trunks, etc., of the vegetation the charring is highest or most intense. The charring is greatest on the windward side at that point, if other considerations are equal. Since unequal accumulation of combustible material may exist, it cannot always be assumed that all other considerations are equal. A number of stems and trunks must be examined before a conclusion is reached as to the direction of travel of the fire in a local area. When the predominant direction is established, further search backward in the fire's direction will ultimately lead to a point of origin. From this point the fire should have spread outward into all areas that are burned. Usually, this point will be on, or very close to, the fire's perimeter; at the foot of a hill, if any is present; or in the interior of the burned area, if the land is relatively level. When a point is found from which the burning has spread in all directions, the origin has been located.

CAUSE OF FIRE

The next step in the investigation of the outdoor fire is to determine the cause when possible. In many instances this will be impossible because the evidence has been destroyed. It is always necessary to make a reasonable search for a cause, and to ascribe such cause to unknown factors only as a last resort. The causes of outdoor fires are well known in general, but, nevertheless, many misapprehensions exist. The common causes are:

1. Unextinguished fires, left by campers, hunters and others.
2. Carelessness with smoking materials, including burning tobacco and matches.
3. Trash burning and, in some areas, controlled grass and brush burning.
4. Sparks from vehicles, especially steam locomotives which are still used in some logging and other similar operations.
5. Lightning, which is a major cause of timber fires.
6. Power transmission lines and accessories such as transformers.
7. Arsonists, who may use a variety of devices but generally use a lighter or match.
8. Firearms, which under certain circumstances may blow sparks of burning powder into dry vegetation.
9. Spontaneous combustion, limited to very specific types of circumstances.
10. Overheated machinery, which may be in contact with combustibles.
11. Miscellaneous objects, sometimes present among trash, including glass that can focus the sun's rays (burning glass effect).
12. Sparks from any source, such as impacts of metal with rock and static discharge.
13. "Sleepers," which are old trees or stumps containing much rotted wood internally in which a smoldering fire has been previously induced. Because of their unique burning characteristics, fire may persist in them for long periods, claimed in some instances to be as much as a year.

It is apparent that many additional types of ignition methods are possible, although rarely encountered. For example, in military operations, certain types of weapons are designed to ignite available fuel. Similar considerations might hold for some blasting operations and almost any similar activity in which heat or fire is utilized. Most of these causes require no discussion here, because their use or effects are matters of common knowledge. A few of them are not as clearly understood and warrant brief comment.

Spontaneous combustion is occasionally claimed when no other source of fire is located. It should be understood that this event is rare, and can occur only under special types of circumstances which are discussed in more detail under "Sources of Ignition." The necessary conditions are much more likely to occur within structures than in the open. However, it has been known to happen in piles of offal, in wet hay, and other moist vegetable material.

Power lines generate fires often enough so that the mere presence of such a line within or close to a fire area renders it suspect as the cause of the fire. Actually, there are a limited number of methods by which a power line may generate a fire. These are:

1. Transformer shortcircuits, perhaps one of the more common and dangerous.
2. Leakage over dirty insulators and support structures when moist, giving rise to "pole top" fires.
3. Arcing between conductors that accidentally come into contact, e.g., in high winds.
4. Fallen wires that arc with the ground or objects on it.
5. Grounding of in-place conductors by external objects such as fallen trees.

The mere presence of a power line in the neighborhood of a fire should never be taken as the cause without a thorough investigation. Too many times, it is easier to blame the obvious than to prove the matter pro or con. Certainly, despite the inherent hazard of power lines as a source of ignition, more fires have been blamed on them than is in accord with the facts.

Sun-kindled fires are rare but not impossible. Any glass object having the general configuration of a lens may focus the sun's rays on combustible material and initiate a fire. Equally, a reflecting material with a concave surface may have the same effect. A globe-shaped bottle or flask filled with a clear liquid is the most dangerous item of this type. Ordinary broken glass will rarely have an appreciable lens, or concave mirror, action. Trash piles especially may yield from time to time objects that have the requisite properties.

The careful investigator who attempts to locate the exact cause of a fire in the exterior must, above all, avoid assuming that, because some hazard existed in the region, it was necessarily the cause. The fact of a fire along a roadside suggests that a careless smoker may have set it, but without finding the residue of the match or cigarette, it would be unwise to do more than consider this as a possible cause, along with sparks from an exhaust or other sources. Conversely, careful search at the origin has at times located a partially burned match or the remains of a cigarette. It also occasionally locates "sets" resulting from the deliberate acts of an arsonist. Most commonly, these are devices constructed from a book of matches and a lighted cigarette. Other types may sometimes be used.

The following examples of investigative experience of the author may be useful to clarify some of the statements above.

1. An investigator located a partially burned match at the origin of a forest fire. Comparison with matches sold in the locality showed it to have a different origin. Laboratory comparisons located the make of the match as that of a small company with a local market in another region of the country. With this knowledge, the investigator approached a person believed to have come from this part of the country, put a cigarette in his mouth, and requested a match. The type of match proffered was the same as that which set the fire, and this case was solved by use of a single burned match.
2. The origin of a grass and brush fire was known to be within a rather restricted area which contained an old trash pile and, closer to a house, another point at which trash was burned. The occupant of the house was accused of illegal burning of trash at the second point, although this was denied. Investigation showed clearly that the involvement of this second point was secondary, since the fire had burned from a lower area uphill to the point at which such trash would have been burned. The origin was close to and possibly at the old trash pile. No actual cause of the fire was found, although several items of suspicious character existed in the old trash pile.
3. A highly destructive fire in a lumber mill was alleged to have originated in a sawdust pile that was within the fire perimeter and claimed to have been ignited some time ahead of the fire and to have smouldered in the interval. Investigation showed that the fire originated in weeds some distance from the sawdust pile, that the fire merely burned over the top of the pile, and that no smouldering fire had ever existed in the pile. Since the entire fire occurred immediately along a railroad right-of-way and a steam locomotive was known to have passed just prior to the outbreak of flames, it is a reasonable assumption that a spark from the locomotive ignited the weeds.
4. A brush fire on a hillside encompassed several branch power lines, the distribution point being within the perimeter. A single witness declared that he had seen a transformer fire on the distribution structure, and claims were made against the power company. Investigation showed that the fire had originated a considerable distance from this structure, burning up the hill toward it. Further, it was shown that no transformer was present either

on the poles or within the perimeter of the fire area. No lines fell until the fire was far along, and no possibility existed that the power lines were causally involved.

5. A power line right-of-way passed immediately downhill from a large mudslide which was held in place by interlacing trees pushed over by the enormous weight of mud above. A fire started close to this right-of-way and burned up this hill. A tree had fallen against a wire, and the top of the tree was heavily burned over a few feet of length. This type of burn is characteristic of electrical conductance, not exterior flame. The allegation was to the effect that the power company was remiss in not controlling the mudslide, even though it was above the right-of-way, and that this had forced the tree down on the wires and started the fire. Investigation showed that the origin was not under or around this tree top but actually further down the hill, below the right-of-way. In addition, it was shown that the fire, by burning away interlocking branches of the trees that supported the mud, had caused the slide to move and force this particular tree against the wires. Thus, the chicken and egg problem was reenacted, with an initial fire causing the grounding of the wire, rather than the grounding of the wire causing the fire. The sequence was established by examining the roots of the fallen tree. Those roots that were originally exposed were all charred, but other roots were also exposed and not charred. These latter roots could have been exposed only after the tree fell and after the fire had passed that point.

9

Sources of Ignition

In connection with specific types of fires, mention has been made of sources of ignition. Without exception, the source is some type of flame, spark, or hot object. In every case the temperature of the source must be in excess of the ignition temperature of the fuel with which this source comes in contact. Further, it is always true that the source is small as compared with the ultimate fire that follows it. These simple principles apply not only to the special types of fire previously discussed but to all ordinary fires. It is true that intermediate stages from initial ignition to final ignition of the destructive fire often exist. The fireplace fire or bonfire may be kindled in the usual manner and later initiate a larger and destructive fire. Likewise, material that is heated in a flame under control may be dropped or blown from its position into fuel which is then ignited. These intermediate occurrences must always be taken into consideration in dealing with the source of ignition of a destructive fire, but they do not strictly violate the simple principles of fire kindling.

PRIMARY IGNITORS

The primary source of ignition of virtually every fire is *heat*. Any other possible source can be disregarded as being so rare and exotic as to have no practical interest. The heat can be applied as an initially small flame from a match or lighter. Without question, this is one of the most important sources of ignition and includes a large proportion of all intentionally set fires. Other sources of heat are from sparks or other electrical discharges such as arcs or lightning. All of these generate enough local heat to exceed the ignition temperature of many items of fuel. Heat is also generated by wires carrying an overload of current. It is generated by numerous chemical reactions of an exothermic type, and it is sometimes obtained in the form of heat radiation of sufficient intensity to initiate combustion. To clarify the relationship of various ignition sources, each will be considered briefly.

Matches

This most common device for kindling a flame is specifically and exclusively designed for this purpose, and it is the most basic source of flame. Whether it directly initiates the destructive fire or merely sets the controlled fire which later is responsible for the larger fire is immaterial to the basic question of ignition, but may be very material to the investigation of the larger fire.

The match is a combination of a stick which is treated to be readily combustible with a head that includes a self-contained fuel and oxidizer, sensitive to friction which generates localized heat. Originally, most matches combined yellow phosphorus and sulfur with a chlorate or similar oxidizer. Yellow phosphorus is now prohibited by law because of the bone diseases it causes. At present, matches are found in two varieties, the "strike anywhere" and the "safety" types. The former contain an oxidant such as potassium chlorate mixed with an oxidizable material such as sulfur or paraffin, a binder such as glue, and some filler such as ground glass. The tip, which ignites first by friction, contains a high percentage of phosphorus sesquisulfide, P_4S_3, which is more readily ignited than the remainder of the head.

The safety match contains an oxidizing substance and an easily oxidizable material such as antimony sulfide. It will ignite only when struck on the box which is coated with red phosphorus, an oxidizing agent, glue, and some abrasive material such as ground glass. The stick of the match is likely to be impregnated with a chemical to suppress glow. Paper matches will generally be heavily impregnated also with paraffin. Numerous variations of match formulation and manufacture exist. The matches are likely to be very different in appearance, and sometimes in action, from one country to the next. Thus, it is sometimes possible to utilize partially consumed matches at the scene of a fire to trace the origin of the ignition material.

In one forest fire, a partially burned match was located near the origin of the fire. This match was different in appearance from any of the matches marketed locally. It was traced to a factory in another part of the country, giving a lead to recent arrivals from that section who were found to have the same type of matches in their possession. This led to the discovery of the person responsible for the fire. It should be noted that paper matches also differ in the nature of the material used to make the cardboard of the stem. An example of this would be inclusions in the paper stock. This also can at times be helpful in elucidating the origin of the fire.

Lighters

Basically, there are two general types of lighter: the electric lighter, which is found in most automobiles, and the liquid-fuel-containing lighter. For most practical purposes, the electric lighter may be disregarded as a source of ignition of serious fires, because it depends on a battery (automobile) to heat the wires to kindle a cigarette. It is not operative except with the battery, which makes its use very restricted. The liquid-fuel lighter generally kindles with a spark from an abrasive and steel. Thus, the basic cause of the ignition is a spark, or hot particle, which ignites the fuel under controlled conditions, leading to a small flame. Such lighters are an obvious substitute for a match, and are, no doubt, responsible for numerous fires. They are rarely left behind and, therefore, are unlikely to be evidential materials. A variation of the lighter is that which kindles by a catalytic oxidation rather than a spark. Here the manner in which the fuel generates heat is different, but the significance to kindling of destructive fires is identical with the more common form of lighter.

Sparks

The definition of spark is ambiguous and, therefore, requires some clarification. Electric sparks are those which represent a discharge of electric current through air or another dielectric. They may at times be associated with the other type of spark which consists of a tiny fragment of burning or glowing solid material that is in movement through the air. This type of spark should be thought of as a hot fragment rather than as a true spark, even though the definition of spark includes both. Both types of spark are likely to be hot enough to kindle fires, especially with vapors and frequently with solid fuels.

The electrical spark is not readily distinguished from the arc. However, the arc persists for a considerable time interval as a discharge, while the spark is generally instantaneous. The arc is obviously more dangerous than the spark because it lasts longer and therefore builds to a higher temperature, allowing more time for transmission of heat to the surroundings. Since they are basically the same type of phenomena, their consideration is otherwise equivalent. Their further discussion is given in a later section covering electrical sources of ignition.

Sparks of the hot particulate variety are often intermediate transmitters of fire from a controlled to an uncontrolled situation. This is their chief function in ignition because, unless there is already a source of heat, such as a small fire, there can be no spark of this type. In this respect they differ from the electrical spark which requires no fire to form it. More extensive discussion of the role of this type of spark in causing destructive fires is considered in appropriate later sections.

Hot Objects

Most hot objects are heated either by being in or close to a fire, or by the flow of electrical current through them. In the automobile cigarette lighter, a current heats the wires. This can happen also in electrical appliances and wiring. Most such objects are therefore wires. In an actual fire, additional materials may be heated and transmit secondary fires by igniting additional fuel that was not burning before. Such fires are not primary but may lead to loss of control of the preexisting fire. They are also discussed later in connection with specific types of fire kindling.

Friction

As a source of fire, this is a special case of "hot objects." Friction between two surfaces generates heat, as in sliding brake bands of the automobile which can become extremely hot. As a source of fire, rubbing two sticks together is still discussed but little used on the modern scene except (traditionally) by Boy Scouts. Even they are more likely to supply a bow to spin a wooden point in a wooden depression because of the much greater speed and frictional heat that is generated. Wood is used, not primarily because it is a fuel, but because it is a very poor conductor of heat. It allows heat to be generated by friction faster than it is carried away and dissipated. With enough effort, the local heat may rise to the point of kindling a glow in fine tinder, which can then be aroused into a flame by fanning or blowing. While this method of kindling fires has historical interest, it is of little or no consequence in the modern age.

On the other hand, friction in other contexts may be of great importance in kindling fires, especially in machinery. The "hot box" on the railroad car has been frictionally heated because of inadequate lubrication, and it may cause adjacent fuel material (e.g., greasy rags) to catch on fire. Any bearing which does not have adequate lubrication can become hot through friction and exposure of the hot object to a readily ignited fuel and lead to a fire. This is without doubt one of the relatively common sources of fire where much machinery is in use.

Not only bearings give rise to dangerous temperatures from excessive friction. High speed rotors, for example, which come into contact with a housing, have been known to produce enough heat locally to melt the housing, as in airplanes. In the presence of flammable vapors or even solid fuels, these temperatures are quite sufficient to ignite a material that is combustible. This possibility must always be considered carefully by the investigator of any fire that seems to have arisen in the neighborhood of a machine in which there are rapidly moving parts.

Radiant Heat

While the heat radiated from a fire may at times kindle other fires at a distance, such kindling is not primary but secondary to an already very large fire. The only primary function of radiant heat, as a source of ignition, is when rays from a very hot source such as the sun are focussed by a lens-shaped object. This will concentrate enough heat to ignite material that was not previously afire. The various aspects of radiant heat were discussed in an earlier section.

Chemical Reactions

A number of chemical mixtures are capable of great heat generation and even formation of flame. They are only of occasional consequence in ordinary investigation of fires because of the relative inaccessibility and frequent lack of knowledge about them on the part of arsonists. They are more likely to be encountered in industrial fires than elsewhere, and they require the attention of the chemical specialist.

THE ROLE OF SERVICES AND APPLIANCES IN STARTING FIRES

So many fires are either started by service facilities or appliances, or are attributed to these origins, that consideration of such possible origins is of the greatest importance. When the origin of a fire is not determined, it is very convenient to decide that it is due to some malfunction of an appliance, a shortcircuit in electrical wiring, or other similar cause. This decision is especially attractive because in many fires, there is so much damage to wiring, or melting of attachments in gas lines, that such an opinion can have some credibility. In addition, it is indisputable that malfunction or failure of services and appliances do cause many fires. Thus, the investigator may readily find himself on the horns of a dilemma in the attempt to separate the real from the claimed origins of fires from such items. In modern building construction, electrical wiring is widely distributed as an integral part of nearly all parts of the structure, and gas lines and appliances are also very common. In all such structures, it is the rare fire indeed that does not engulf some such items to some extent at least, thus possibly focussing attention on them.

Gas Lines

Because of the properties of gas, outlined elsewhere in this volume, it is clear that gas enclosed in a pipe does not pose any hazard. Unmixed with appropriate quantities of air, the gas is not flammable. Only when it escapes is there any danger associated with it, because then it can mix with air to form a combustible or explosive mixture. Such an escape is possible because of: (1) leakage due to inadequately sealed joints or corroded pipes, (2) mechanical fracture of lines from external causes, and (3) failure of lines because of excessive heat which may melt essential sealing components of the line or gas meter. Each of these requires separate consideration.

Leakage. A gas line may show leakage from inadequate sealing of the joints during installation, from deterioration of joints, or from corrosion which is severe enough to penetrate the walls of the line. It may also occur from worn or defective valves and fittings. In any of these events, it is expected that the odorous material intentionally added to all natural gas will serve as a warning of escaping gas. Natural and some manufactured gases are essentially odorless. Needless to say, in an empty structure or in the exterior, there may be no way of discovering such a leak even with an odor constituent present.

Assuming that such a leak exists and that it is not noted, the sequence of events is as follows: Initially, the gas is too low in concentration to be ignited from any source. As it reaches the lower combustible limit, it will have formed an explosive mixture throughout all of the area in which the gas concentration has reached this limit. This will normally occur only in a confined or partially confined space, but that space will be under great danger of an explosion—not a fire, initially at least. If there is a source of combustion, such as a pilot flame, or if a match is ignited by a smoker at this time, an explosion will occur. Often it will be very forceful, which is characteristic of a lean mixture, and without continuing flame. If it is not ignited while still a lean mixture but accumulates to form a rich mixture, it will generate proportionately less mechanical force but will hold flame after the explosion. *Unless ignited immediately at the leak*, where there may be locally a small amount of gas-air mixture within the combustible limits, *only an explosion, not a fire, is expected as the primary event.* In most gas leakage, the concentration never reaches the explosive or combustible limit, and no fire or explosion results.

Mechanical fracture. Release of gas from a mechanically fractured line is not expected to occur too often. However, it is probably far more common than is generally believed, and occurs even in the absence of natural catastrophes such as earthquakes, landslides, and similar major causes of destruction. For example, a gas line was laid under a street over which there was heavy trucking. Failure of the workman to tamp properly the soil around the pipe allowed it to buckle under the recurrent loads until the line parted, filling the soil with gas which eventually led to an explosion and fire. In another instance, a small gas stove was attached with a copper tube to the gas line and with a soldered connection. The stove was moved around somewhat with recurrent flexure on the connection. Finally the joint parted and resulted in both a fire and a multiple asphyxiation by the combustion products.

As with leakage, release of gas under these circumstances is likely to lead to explosion, and will cause fire directly only when the gas is ignited as it first emerges from the fractured line. This is what happened in the second illustration above. Since the small heater was burning, the gas that escaped from the opening in the joint was ignited before it could mix with enough air to form an explosive mixture.

Failure from heat. In general, a failure of a gas line from excessive heat will be limited to a general fire with which the gas is not causally related. Lines are frequently joined by low melting alloys such as solder; if such a joint is heated by exterior flames or even hot gases, the joint may fail. In very hot building fires, it is not uncommon to have brass or bronze fittings partially melt. Since these are regularly used in connection with gas lines, such mechanical failures will be encountered.

Because there must already be a fire to melt the attachment in these situations, it is clear that the escaping gas will merely feed additional fuel to the fire *locally* and will produce a characteristic "blow-torch" type of burning ahead of the opening. It is not to be considered as a cause of the fire, but only as a result and a contributing factor to the intensity of the fire. *The gas cannot spread under these conditions to augment the flames except ahead of the line opening.* Under no conceivable circumstances will gas escaping into a burning fire be consumed at a distance from the opening, but it will at that point create a large "blow-torch" type of flame. This will have the effect of causing a burned out area immediately ahead of the opening. Its effect on the remainder of the fire will be to all intents and purposes negligible in virtually every circumstance. It may alter the fire pattern slightly because of local intensity which tends to deflect air currents to the hottest part of the fire.

Despite these rather obvious facts, it is quite common for claims to be made that escaping gas augments severely the total fire throughout a building. This necessitates that the unburned gas pass through, around, or over burning areas to reach distant portions of the fire, which is palpably ridiculous.

Gas Appliances

Appliances used for purposes other than industrial are relatively limited to a variety of heating devices: stoves, ranges, room heaters, furnaces, clothes driers, and water heaters. All of these items have similar principles. Thus, regardless of the exact type of heating device, the investigations, as well as all the possible hazards that attend the use of the device, are very much the same. The basic differences that exist are much more likely to be significant in asphyxiations than in fires. For these reasons it is not generally of great concern to the fire investigator what may be the BTU rating of the appliance, whether the flame is adjusted correctly, or whether the automatic cut-off mechanism is functioning. The latter may be important, but generally because of an explosion rather than a normal fire.

One of the few ways in which the gas appliance is likely to initiate a fire results from the simultaneous failure of the thermostat and the high-level control. This device is a second thermostat located in the circulating air ducting of the furnace. It is usually coupled in series with the regular thermostat and is adjusted to break the circuit at a maximum temperature above which the furnace cannot be safely operated. This temperature varies but is usually above 200°F. A simultaneous failure of both thermostats allows the furnace to continue operation indefinitely with great production of heat and often a resulting fire.

Although gas appliances do start fires, it is believed that they are probably less often the cause of a fire than is generally assumed. Gas is the chief combustible in many places, and its intrinsic hazards tend to build an abnormal fear of it as a source of fire. It also serves as a convenient "probable cause" when the investigator finds no other that is provable.

When a gas appliance starts the fire, the fire pattern itself will often, if not usually, reveal the fact. Associated with the pattern which shows the origin to be behind, around, or over the gas appliance, there must invariably be some malfunction, improper installation, or other detectable fault or defect. Only the combination of factors can be considered as proof of origin from a gas appliance. The reason for the double requirement is relatively obvious. The arsonist often will set his fire under or close to a gas appliance in order to make it appear that the appliance is responsible for the fire. When this is done, liquid flammables are likely to be poured around the appliance, and the fire started so that the origin seems to be obviously associated with the appliance; an incorrect inference will result. In such an instance, such flammables are likely to penetrate below the appliance and possibly below the floor, and produce the characteristic charring and damage in a region at which a gas fire could not have reached. Gas fires are more subject than any other type to burning upward, since the fuel as well as the flame is significantly lighter than the surrounding air.

All approved gas appliances are surrounded by some type of insulation. In circulating heaters, this insulation consists merely of an outer wall separated from the wall of the firebox, through which circulates the cool air that is to be heated. Such a wall will not become significantly hotter than the ambient air that circulates behind it. Presence of flammable trash against such an appliance wall will not lead to ignition of the trash. The same cannot be said of the wall of the combustion chamber which may well be hot enough to ignite material against it. However, only a very unusual circumstance would cause such a contact.

A common claim that there has been a backfire from a furnace or other heating appliance carries somewhat more credibility. Nevertheless, it is an extremely rare happening which can only be the result of gross maladjustment within the appliance. To obtain such a "backfire," it is necessary for the gas to accumulate in some quantity in the combustion chamber before it is ignited. Upon ignition, a small explosion occurs and may blow flames out of the open front of the appliance. If there is a good combustible in the path of this small amount of flame, a fire could result. The combustion is so unlikely, however, that any claim of this type must be very carefully scrutinized before being accepted. With a pilot light, it is difficult to accumulate sufficient gas for any significant backfire, although maladjustment of some type may be effective. In any event, so gross a defect is not difficult to recognize. In the absence of a pilot light, the gas would not ignite until it diffused or spread to some other source of ignition, which in all probability would require enough gas to produce a violent explosion. In addition, escaped gas will often pass harmlessly up the vent pipe into a chimney. It is difficult to imagine any reasonably normal situation in which fires are kindled by this type of situation.

Another manner in which a gas appliance can start a fire, and would be expected to do so, involves direct contact of a gas flame with flammable or combustible material. In general, this will occur only with open flames, although instances of flammable materials being drawn into a protected flame will occur. Open flames are most common on kitchen ranges, laboratory and industrial burners, and special types of heating appliances. Corresponding to this fact, such items are more hazardous as sources of fire. A very common example is the ignition of cooking oils and greases in the kitchen, not to mention such things as carelessly dropped rags, paper, and the like. Greases are especially a cause of fires in restaurants where the amount of cooking is great and time spent in cleaning of stove tops may be minimal. Such fires may also spread upward in grease-coated vents and carry the fire up with a great increase in its magnitude.

Small gas appliances that are not part of a permanent installation can be displaced or upset with occasional drastic results. Laboratory burners are especially subject to this type of event. They can start fires by playing the flame on top of a wooden table, for example, or against some other flammable material. Similar considerations may hold for small portable gas heaters sometimes used in homes and similar areas. Note should be made that defective burners in installations of a permanent or semipermanent nature may at times become subject to similar considerations. When this type of possibility exists, it should be recognized by the investigator.

An unlikely, but possible type of fire causation from a gas appliance involves some alteration in the gas nozzle in the venturi, possibly by partial or total unscrewing of the nozzle. If this should happen, much larger than normal quantities of gas would reach the flame which, in turn, might become much bigger and spill from the top or front of the appliance. The author has not witnessed such an accident, but it is conceivable, especially when tampering or ignorant efforts at repair or adjustment have occurred. Such fires would necessarily occur at the first use of the appliance after the opening had been enlarged, which should again make their detection and proof of cause relatively simple.

LP Gas

In addition to natural (or manufactured) gas which is common in urban areas, many rural and isolated regions make use of the so-called LP (liquid petroleum) gases which are purchased in tanks. These are normally either propane, butane, or a mixture of the two with small quantities of other similar hydrocarbons. Because both propane and butane, especially the latter, are heavier than air, their escape into the air has features somewhat different from that of methane which is a major constituent of natural gas. Being heavy, they behave somewhat more like the fumes of gasoline, the effects being described under "Properties of Non-solid Fuels."

It is of interest that the LP gases apparently give rise to fires (and asphyxiations) somewhat more frequently than is true of natural gas. The reasons have little if anything to do with the nature of the material, but rather with the method of installation of the necessary storage and distribution equipment. Whereas natural gas normally is dispensed from heavy iron pipe, permanently installed, and generally under control of ordinances, the LP gases are ordinarily led around in semiflexible copper pipe and in a relatively uncontrolled manner, sometimes by amateur installers. In addition, the supply of gas has to be replenished from tank trucks or with portable tanks rather than from a permanent installation. Thus, there is a considerably increased probability of poor connections, broken copper tubing, leaking valves, and similar deficiencies. It is these deficiencies, rather than any inherent property of the fuel gas, that leads to fire, explosion, and asphyxiation hazards.

In investigating fires that are in connection with LP gas, the investigator must take special pains to check all lines, valves, and connections. The fire pattern itself will give a very good indication as to the portion of the distribution system that was at fault, so that finding the defect is ordinarily not difficult. In the differences noted in delivering the gas to the appliance, or possibly in the appliance itself, will reside the only reasons for any special hazard that may attach to the use of these gases.

Electric Wiring

Electricity, like gas, carries potential danger of causing fire, but unlike gas, the type of effects and the necessary investigation are quite different. There are several fundamental principles to bear in mind which will prevent most errors in assigning fires to electrical facilities. All of these apply to total circuits which must include all electric appliances along with the wire that brings them power.

1. *Heat and, therefore, fire can only result from a closed circuit.* As long as any circuit remains open, there is no possibility of it kindling a fire. Thus, closure of a circuit that is not normal to the functioning of the facility is a feature of the investigation that must be sought. At the same time, it must be realized that the current in every part of the circuit is also that in the total circuit.
2. *The current flow in all circuits is controlled by Ohm's Law,*

$$I = \frac{E}{R}$$

where I is the current (amperes), E the potential (voltage), and R the resistance, generally expressed in ohms. All conductors of electricity have resistance, including the lines that supply current to appliances. The resistance of these lines should always be much lower than that of the resistances of the appliances attached to them. If this is not the case, the lines will heat and present a hazard. When the resistance in the total circuit is drastically reduced, as occurs in a shortcircuit, a very large surge of current through the circuit will result.
3. The heat generated by an electrical circuit is in direct proportion to the resistance, to the square of the current, and to the time, expressed as

$$\text{heat} = kI^2Rt$$

where k is a proportionality factor to reduce the heat to some standard unit, such as calories or British Thermal Units. Thus, in the shortcircuit, the total resistance falls and produces a surge of current, but major generation of heat is in that part of the circuit where the resistance is greatest.

In order to make calculations on the basis of the above equations, all units must be defined. Ordinarily, secondary equations derived from the above fundamental equations are used. However, it is rarely necessary for the fire investigator to do such calculations. Rather, it is for him to understand the principles, so that he will be in a position to approach the remains of electrical equipment in the burned area intelligently and not be misled by effects that he is unable to evaluate.

Electrical wiring. Along with an understanding of the basic physical laws that relate to the behavior and properties of an electrical current, the investigator also requires some knowledge of the methods of constructing electrical circuits in structures, so that he may interpret better what he sees. Elementary knowledge, at least of the principles of furnishing electrical service, will be basic to the necessary understanding.

It was noted above that the heat generation from a current is proportional to the square power of the current passed through the conductor. In long-distance transmission, this loss of power through dissipation of energy as heat is economically detrimental. After all, it is the watts delivered to the customer that are paid for. The number of watts is expressed as the product of the voltage times the amperage, and it is a measure of the actual power available to the customer. The formula is

$$V \times A = W.$$

Volts deliver the potential, amperes deliver the heat, and in a loose sense, watts deliver the total power. Thus, in the long-distance transmission line, it is advantageous to maintain a very high voltage, because this can deliver power with minimum heat loss in the transmission. When delivered to the customer, it is desirable to provide a lower voltage which is much safer. Because of the equation above, a given quantity of total power at low voltage gives a much higher amperage, which is related to heat developed, as well as to the power of the motor that is operated.

With this fundamental fact of delivery of electrical current to the household, it remains to describe the method whereby the current is made available to the householder, in order that intelligent study of the relation of current to the initiation of fires can be better understood. It should be realized that in various portions of the world, current is delivered somewhat differently. Some places depend on direct rather than alternating current. The voltage available to the household varies as well as the cycles of alternation. However, in most of the United States, alternating current at about 110–120 volts and 60 cycles (oscillations per second) is the rule. Ordinarily this is made available in a three-wire system which is illustrated in the figure. A much higher voltage, frequently 12,000 (12 KV), is brought to the region of delivery. This current is fed to a step-down transformer, which reduces the total voltage to 220–240 volts, with a correspondingly higher available amperage for any given wattage or total power factor. The figure shows the high voltage input led to a transformer which makes the transformation in voltage (and available amperage). From the transformer normally come three wires. Two of these actually carry the voltage, generally 220–240 volts, the third being a neutral or ground wire carrying no voltage but at ground potential. The two wires that are charged may be termed "hot wires," and a connection between them will have a total voltage of 220–240 volts. If either is connected to the ground or neutral wire, the voltage will be just half of this value or 110–120 volts. This is because the actual output from the secondary circuit of the transformer is attached to the hot wires, the ground wire being tapped at the center and connected with the ground literally. Metallic conduit, in which wiring is frequently placed is also at ground potential. Direct grounding of these various items in connection with the installation of structural wiring is the rule.

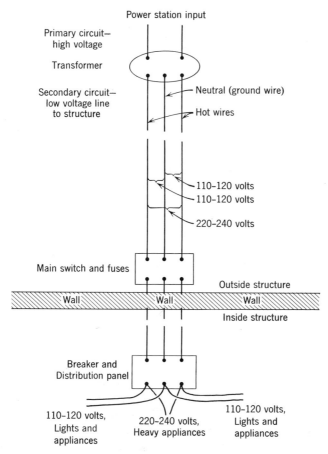

Figure 2. *Schematic of common three wire system for supplying electric current.*

It will be noted that the transformer drops the voltage of the incoming line to the potentials available for the final house circuits. This secondary voltage arrives at the structure and is immediately led into a metal box 1, normally on the exterior, containing a main switch that turns off or on all house current. It also contains heavy fuses suitable for whatever load may be expected in the entire structure. A line is led (properly in conduit) to a second box 2, containing the distribution panel in which the single circuit is broken into a number of secondary circuits that lead to various points of the structure. Each of these circuits carries a fuse or breaker usually set for 15 or 20 amperes to protect the individual circuit. Ordinarily, all outgoing secondary circuits are 110–120 volts rather than any of them being 220–240 volts which often utilizes a separate box. The individual circuits are rarely carried in rigid conduit, nor is the wire as heavy as in the single main circuit wiring. It must not be forgotten that however many branch circuits there are, all of them are a part of the total circuit, and effects on any portion of that circuit may affect all. For example, a shortcircuit that interrupts the incoming single circuit prevents current from reaching the lateral branches, and a heavy shortcircuit in a lateral may also produce destructive effects on the main single portion.

Shortcircuit possibilities result when any two elements of the conducting environment are at different potentials and may come into contact with each other. The various possibilities may be outlined as follows.

1. High tension wires bringing current to the transformer may contact. This will almost invariably be in the open and will result in heavy arcing and fusion of the lines which will often separate by being melted.
2. Either high tension line and ground may contact. This occurrence results in heavy arcing with the ground and is extremely dangerous to anyone in the vicinity when it happens.
3. Coils in either the primary or secondary circuits of the transformer may allow contact, resulting in a transformer fire and destruction. Voltages involved may be variable, depending on the portions in which such short-circuiting occurs.
4. Hot wires of the secondary circuits outside or inside of the structure may come into contact, producing less heavy damage because of the lower 220–240 volt potential.
5. Either hot line may contact ground. This is the most common type of shortcircuit. By ground in this instance is meant a variety of objects including the neutral wire itself, the conduit, any conducting object that is itself grounded through a water or gas pipe, the human body in a bathtub of water, most appliances, and a variety of other objects. Possibilities include such a situation as damaged insulation inside of a conduit, in which either hot wire may make contact with the ground wire, the conduit, or the other hot wire. Shortcircuits in conduit, although possible, are rare.

All circuits in a building are normally installed so that contact between wires that could produce a shortcircuit is impossible. This is often done by use of physical separation between the wires, which are supported at short intervals to avoid accidental contact. On passing through wood or other structural material, such wires are heavily insulated to avoid a circuit being formed through the structural material and causing, to some degree, a shortcircuit. Alternatively, and more commonly in newer structures, both wires of the circuit are contained in heavily insulated, flexible cable, or insulated wires are carried inside rigid conduit. Many varieties of wiring materials are manufactured and used, all of which are subject to local wiring regulations, and most of which have been tested and approved by the Underwriter's Laboratories. Deviations from these practices may be encountered when householders or other amateur electricians install building wiring. These are especially subject to the attention of the investigator.

In addition to the details that must be determined in installation of the wiring itself, there is always present some protective device, either fuses or circuit breakers close to the entrance of the wires from exterior lines. Their function is to break the circuit if an overload that generates too much heat is accidentally established. Both a normal appliance overload and a shortcircuit which produces a surge of current should either "blow" a fuse or actuate the circuit breaker, thus diminishing the time factor in the above heat equation and protect against fire.

It is frequently noted that because circuit breakers have a finite time of response, serious heating effects may be produced before they shut off the current. Probably a more serious situation is due to the fact that the fuse or circuit breaker alike must always be able to handle a brief overload. In starting motors, for example, such an initial overload is normal but limited in time. Thus, a protective device rated at 20 amperes will shortly discontinue a current in excess of this value. An arc may well be created locally that consumes less than 20 amperes, and will not cause the circuit to break until after a fire is initiated. This source of difficulty is most apparent with the circuit breakers or fuses that protect the total current to an installation. Here, the amperage may be set at 60 amperes, for example, because the total load expected may approximate this value. A load of 20 amperes caused in a secondary circuit would not affect the main circuit breakers at all, but could well start a fire locally.

Any fire that results from overheated wiring of the building will ordinarily derive from some overload that heats the wires unduly and is not terminated by the protective device. Direct shorting of wires in the permanent installation is very rare, and can only occur under highly unusual circumstances. One of these, naturally, is a preexisting fire which may destroy insulation and allow contact of wires, which are in no way at fault in starting the fire.

In every conduit, flexible or rigid, there will be two or more insulated wires lying side by side. When the conduit is heated from an exterior fire, the insulation may melt, pyrolyze, or burn, and allow contact of the metallic wires with a resulting shortcircuit and great production of heat. Frequently, such wires will be melted off or fused together. One of the most prevalent mistakes made by investigators is to assume that the fire was therefore of electrical origin, because evidence of a shortcircuit is found. Inside such a conduit, it is unlikely that any shortcircuit that does occur will set a building on fire, but rather that the exterior fire will cause the shortcircuit. Similar consideration also applies to electrical cords attached to appliances and, in fact, to any situation in which insulated wires are in close proximity and separated only by insulation which is subject to damage or destruction by heat.

Appliance wiring. Attachment cords of appliances are a frequent source of deficiencies that may initiate fires. These are probably the most important of the electrical causes of fire. Not only are such cords less well insulated, both electrically and mechanically, than is true of the better installations of permanent types of wiring, but they are subjected to many kinds of use and abuse.

Only two basic difficulties normally arise with appliance cords. They are heated either by an electrical overload or from some external source, and the plastic insulation softens enough so that the wires can contact and cause shortcircuiting. Faulty construction, generally in the plug attachment, allows shortcircuiting. Naturally, mechanical damage which breaks insulation may also allow contact of wires. Many cords are allowed to become so old that the flexing and bending of the cord finally allows deterioration both of the wire and especially of the insulation. Such cords, near to breaking, are distinctly hazardous in producing shortcircuits.

In the installation of permanent wiring of buildings, one of the necessary operations is drawing the wires through the conduit, which frequently takes the wire around a rather sharp bend. If the bend is not smooth or the insulation of the wires tough enough to withstand this operation, insulation may strip inside the conduit. Ordinary temperature changes, vibrations, and similar effects can then cause shortcircuits within such a defective wiring system within the conduit. Such an event may conceivably cause a fire but is not likely to do so, because the protection of the conduit itself tends to dissipate the heat. From the investigative standpoint, the issue is very clear. If such a shortcircuit should occur within the conduit, all circuits past this point are inactivated, and no shorting of lateral circuits is possible. If such shorting has occurred in secondary circuits, as it does in many fires, the possibility of a shortcircuit within the conduit is ruled out as the cause of the fire. This error has even been made in the examination of fires by electrical engineers who should be aware of the properties of circuits as a source of fire.

Wiring within an appliance, as opposed to the cord, is far less subject to damage from anything but natural wear and tear. *Heating coils* gradually deteriorate from oxidation and may lead to shorting, or more likely a wire will finally burn through with formation of a local arc when separation occurs. Such units are usually immune to starting a fire because, although their normal state in use is hot, they are not in conjunction with anything flammable. If they were, the heat they normally generate would also start the fire.

Electric motors probably cause more difficulty than other appliance components. They not only contain wires, but some of these are in motion, and there is often a commutator which is subject to sparking. Dirt and foreign objects may get in them and damage insulation, wires, or other components. Except for induction motors, volatile flammables must be rigidly excluded, or a fire or explosion is almost inevitable from spark ignition of the gas. One of the common difficulties with motors results from frozen bearings. While running, self-induction produces a potential counter to the current source, which prevents large currents from flowing through the motor. When stopped for any reason for any length of time, full current flows, controlled only by the total resistance of the motor's coils. This current is usually enough that in a short time it will seriously overheat the motor. Furthermore, it may not exceed the capacity of the fuses or breakers in the line, because these must always have a capacity sufficient to deliver this amount of current for starting.

In appliance fires, or those so suspected, the major item for examination is ordinarily the motor. Its role should be apparent from the pattern, either of burning or heat damage such as discoloration of metal. It will often also be indicated by fused metal, within the motor itself, even when the final fire would not be expected to melt the metal in question.

So many special types of appliances are in use and electrically operated or controlled, that it is not possible to cover all of the possibilities in a short discussion. In all such instances, careful examination of the remains of the appliance, along with consideration of the fire pattern, should be successful in locating the origin of the fire. In some instances the damage is so great as to effectively obliterate the nature of the fault. Such a situation may, but should not, tempt the investigator to attribute the cause to the appliance solely because of the extent of damage. In numerous instances, a preexisting fire has led to shortcircuiting which caused the effects observed, without having any causal relation to the fire. When the origin of a fire as indicated by its pattern is immediately adjacent to fused wires, and there is evidence of great local heat, it is highly probable that the fire was caused by the shortcircuit. In this event, it is necessary to determine, if possible, why the shortcircuit occurred. If this can be done, there is unassailable evidence as to the electrical origin.

ROLE OF HOT AND BURNING FRAGMENTS IN KINDLING FIRES

Hot and/or burning fragments, commonly termed "sparks," are in a special category in regard to kindling of fires. As pointed out elsewhere, the designation as "sparks" is an ambiguous one and can scarcely be extended to include bits of hot metal, even in its common context. Special consideration of this type of material is necessary because it is not generally a primary source of ignition but a secondary one that traces back to the origin of the hot fragment in another environment, usually another fire. Most such fragments are glowing, although they may at times be flaming. Bits of wood, paper, and other light organic materials are most susceptible to formation of glowing fragments, and larger pieces of paper may be traveling while still flaming. Bits of paper or tobacco may become detached during the lighting of a cigarette or pipe. These naturally have the propensity of igniting other fuels with which they come in contact. It should be remembered that most such fragments will be carried upward by the draft from the fire that generates them, and air currents may carry them for significant distances into other fuel that is not yet alight.

A question frequently arises as to how far such "sparks" can travel and still cause a secondary fire. Although no precise answer is possible, some principles can be stated. The ambient wind is highly important, because such "sparks" tend only to rise with the air stream over the fire, drift a short distance, and slowly fall unless propelled by wind. Another factor is the type of material which is burning. Very small fragments and thin materials such as paper will normally be totally consumed before they return to some flammable surface. Fragments of wood, excelsior, or corrugated cardboard will burn considerably longer and travel farther before igniting something else. For the latter, distances of about forty feet have been observed. Light items will probably only rarely travel as much as twenty feet, except under special wind conditions. Another factor is the height which they reach before being blown to one side. The higher they go, the more time they have to burn up totally and the longer is the exposure to any air movement. In any event, distances in excess of thirty to forty feet are to be accepted only with the greatest caution.

Throughout this discussion of spark travel, the role of wind has been repeatedly referred to as modifying the situation. No investigation involving the distance and/or effectiveness of such a spark is complete without ascertaining from meteorological records, or at least from witnesses who were present, the wind conditions of the time in question. It is evident that violent winds will produce very different effects from mild ones. However, the difference is not that which might be supposed. A strong wind will blow a spark farther than a weak one, but it will also speed the rate of combustion. With a given size of burning fragment, the effect is not greatly varied with different wind velocities. The most significant effect of the strong wind, by far, is its ability to blow larger fragments of material, which will naturally burn longer than will small ones. A wad of burning newspaper on the ground will not ordinarily travel along the ground for any significant distance if propelled only by a breeze. In a strong wind, however, it may roll for a considerable distance before it is consumed. The relation of importance is the time necessary for it to burn as compared with the speed with which it is moving laterally with the wind. The consideration is far more relevant to materials at ground level than to those that emerge from chimneys, because very large burning objects are rarely carried up a chimney to any significant height.

One consideration of the effect of wind on chimney action is also relevant. Wind blowing across a chimney opening exerts what is known as the Bernoulli effect of lowering the pressure at the chimney mouth. This causes a stronger draught up the chimney which can increase the size of fragment that can move up to the top and be ejected in the stronger wind stream. This adds a further reason for the investigator to be very careful in his consideration of wind velocity when there is a possibility of spark movement having started a fire.

There are several common origins for hot fragments, and since these are the primary source of the fire that may result, they will be considered individually.

Fireplaces and Chimneys

The fireplace especially, but some other chimney structures as well, is blamed for initiating many fires, some of which it undoubtedly does. Nevertheless, the mere fact that a possible fire origin is close to a fireplace is not to be considered as positive indication that the fire was initiated from this source. Four ways in which a fireplace is instrumental in initiating a fire, are as follows:

1. Emission of sparks or blazing materials out the front or open portion, which may burn the surface of floors but will only start a major fire when something more flammable than a floor surface is contacted by the ignited material.
2. Emission of sparks or blazing material from the chimney which then may fall on roofs or other flammable materials and initiate fires.
3. Emission of sparks or blazing materials through defects or holes in a chimney so as to ignite timbers or other construction material adjacent to the chimney.
4. Overheating of flammable materials in the neighborhood of the fireplace or chimney.

It is clear that the second possibility is the most dangerous from the standpoint of kindling large fires. It is also clear that a large proportion of them will be roof fires, although blazing residues or large glowing sparks may travel some distance and kindle fires elsewhere than on a roof.

Chimney fires, which are the greatest danger factor, arise because soot, dust, cobwebs, and a variety of flammable materials sometimes are allowed to accumulate in the chimney where they may be ignited by a spark from the fire below and lead to a large blaze within the chimney itself. When this happens, large quantities of burning fragments will be expelled from the chimney top and greatly increase the danger of igniting a roof or other adjacent combustible structure. Unless this happens, the chimney fire is not by itself especially hazardous. Because of the characteristic pattern and behavior of a roof fire, as discussed elsewhere in this volume, there will rarely be a serious problem in interpreting the cause of the fire.

Serious fires from cause No. 1 are rare, because the top of a floor is very difficult to ignite. Scorch marks are not uncommon and are annoying, but are not subject to general fire investigation. Some rugs or carpets are sufficiently flammable to be ignited under these circumstances, and occasionally furniture may be ignited. Thus, there is a hazard from this source, but again it is relatively minor and readily interpreted when encountered.

Fires kindled by the third method are equally rare but may be very serious when they occur. Few chimneys become so defective by use or abuse as to pose any real danger of leaking sparks or burning materials through their walls, but it can happen as a result of anything that causes serious deterioration of the house, such as earth settling or slippage, earthquakes, or extreme age and decrepitude. Since the chimney is not likely to be destroyed by the fire, the evidence of gaps in the walls, added to the fire pattern, should, upon investigation, establish the origin of the fire satisfactorily.

Method No. 4 is probably the least important of the ways in which a fireplace may kindle a larger fire. However, fires in the immediate neighborhood of chimneys are not too uncommon, and many persons attribute most of these to overheating of the structure around the chimney. Similar considerations apply to furnace and waterheater venting when it passes through openings in wooden structures. It must be remembered that in all such instances, there is a chimneylike structure surrounding the actual chimney or vent. Any fire that starts from any cause in the vicinity of such a structure will be drawn up it with localization of the fire in the region around the suspected chimney or vent. For this, and other reasons, such a postulated origin of fire should be scrutinized most carefully before accepting it. Some considerations of the properties of fuels are relevant.

Wood, especially in massive form, requires a considerable amount of heat to cause its ignition. In fact there must be a local heat sufficient to distill volatiles from the wood before any flame can appear, and the temperature must reach the ignition temperature of these volatiles. Temperatures in this range are not expected to occur around chimneys under any ordinary set of circumstances. It is even more difficult to justify the existence of a smouldering fire, because this normally only follows the first flaming stage and will, therefore, require even more drastic circumstances. Admitting that dry wood is more readily ignited than wood with higher water content, and that recurrent exposure to raised temperature will dry the wood, the requirements for its ignition are not altered basically, but only to the extent of lowering somewhat the required ignition temperature. The works of McNaughton (1) and Graf (2) do not indicate increased hazard as the wood dries, except when the temperature reaches a value around 275° to 280°C. (527° to 536°F.). Such temperatures may be reached in unusual instances around chimneys but are not to be expected. In the absence of any evidence for lower ignition temperatures of dried wood, it is well to consider reports of such fires with caution before accepting the theory of an overheated chimney.

Trash Burners and Bonfires

As sources of hot and burning fragments, trash burners and incinerators rate high. The types of material normally destroyed by burning in such devices are those most likely to form burning fragments, and the necessity of openings in the incinerator to allow ventilation of the fire also allows escape of burning fragments. Most such devices have little or no chimney associated with them, thus minimizing the rise of escaped fragments and, therefore, diminishing the area of hazard from them. Thus, it would be uncommon for a roof to be ignited from a trash burner, but dry grass, for example, in the neighborhood of the burner may well be set afire. On the other hand, a high chimney increases the time allowed for combustion of fragments, so that the hazard is reduced to the degree that the emitted materials are less likely to be actively burning when they emerge from the chimney. It is fortnate that fires started by trash burners will rarely require much investigation because of the obvious nature of the origin and the fact that the pattern will trace directly to the burner in most instances.

The bonfire poses a somewhat different problem than the trash burner, despite the fact that it also is an exterior fire. To maintain a satisfactory bonfire, it is customary to use relatively massive wood instead of paper, packing, cardboard, and similar items of trash. The wood poses relatively much less hazard from rising fragments of burning material than does trash. On the other hand, there is no restraint or protection of the fire from winds, as is true of the trash burner. In a high wind, a bonfire becomes very dangerous from the standpoint of kindling larger, uncontrolled fires. If it is adjacent to dry grass or leaves, it is always hazardous. Here also, the investigation is likely to be rather simple. An occasional difficulty is in distinguishing the remains of a small bonfire from those of a general fire that has involved a concentration of fuel at a point which may leave excessive ash and charcoal which appear similar to the remains of a bonfire. Careful examination will usually distinguish the two. In many instances the remains of bonfires contain, in addition to fuel remains, those of food, containers, or other indication of human activity.

Metals

As a direct source of ignition of fires, metals are of minor importance only. Molten metal may carry enough heat to ignite susceptible fuels with which it comes in contact and is, therefore, capable of starting fires to this degree. Since it requires a considerable amount of heat to melt the metal, such dangers are associated substantially only with special operations in which metals are heated, as in welding, soldering, cutting, and certain industrial operations in metal fabrication. Places in which such operations are carried out are more than normally susceptible to fires from this type of ignition. However, the dangers are often exaggerated, since many alloys, including some types of solder, will not generally carry enough heat to ignite fuel material in which they fall (See Appendix 1).

Molten ferrous alloys, such as steel, are far more dangerous than solders. Thus, welding and cutting of steel with an arc or torch must never be done in the presence of susceptible organic fuel materials. By susceptible, it should be understood that the state of subdivision of solid materials is critical in determining the susceptibility. It is, for example, almost impossible to ignite a large timber accidentally by the short, direct application of a torch flame, or of molten metal from it. Small debris, on the other hand, such as packing, sawdust, or paper may very well be ignited under the same circumstances.

Another origin of molten metal is from fusion resulting from the heat of a shortcircuit in electrical equipment. While hazard from dripping of such material from the equipment is certainly present, the greatest danger is generally from direct ignition by the shortcircuit itself. Thus the metal is to be considered as a secondary and less probable source of ignition.

In considering hot or molten metal fragments as sources of ignition, it must always be kept in mind that most metals have a high density, and the hot fragment will fall rapidly, in contrast to bits of paper or wood which tend to rise in the air stream of the fire itself. Aluminum and magnesium are both of relatively low density; in flat form, e.g., a roof section, they may rise rather than fall. This will not be true of solder, molten brass, iron, or copper. It should also be remembered that metals can burn, although most metals are extremely difficult to ignite. Again, magnesium is the chief exception, with aluminum also being comparatively easier to ignite than other metals.

Hot Fragments from Ammunition

Forest and grass fires especially, and structural fires occasionally, are ignited by residues of ammunition in shooting. Although the hazard from burning powder is extreme only with improper weapons or ammunition or both, there is always some possibility of such burning fragments causing ignition of dry fuel. This possibility is one that the investigator must also consider in assigning a cause to exterior fires and, under special circumstances, to interior fires.

More dangerous sources of ammunition fires are tracer and incendiary bullets, normally used only in military weapons. Incendiary bullets carry white phosphorous which burns spontaneously with great vigor. Ammunition such as this should never be fired where a fire hazard is to be avoided. It could be fired deliberately by arsonists or accidentally by persons who come into possession of this type of ammunition unwittingly without recognizing it. There is no fundamental distinction between hot or burning fragments from ammunition and those from any of the other sources discussed.

The powders used in ammunition all burn at a rate considerably below that of the high explosives, but very much faster than fragments that depend on the oxygen of the air for their combustion. Thus, most powders will actually burn little or none past the muzzle of the gun barrel, although there are exceptions to this generality. In any event, the hazard is diminished to essentially zero when combustible material is more than two or three feet ahead of the muzzle of the weapon. This represents the maximum distance that most guns can throw even molten powder. In most instances, normal weapons, loaded with normal powder, do not throw any appreciable flame past the muzzle.

SMOKING AS A FIRE ORIGIN

The widespread prevalence of smoking, as well as the general opinion as to its causation of fires, calls for some special comments because there are many misapprehensions relative to this matter. There is no question that many fires are started by smokers, both indoors and outdoors. Nor is there any question that the carelessness of the smoker is the primary reason for this high incidence. On the other hand, it is common to hear theories about how smoking caused a fire when the theory is totally inappropriate and the alleged ignition would not be successful under the circumstances.

Smoking involves both burning tobacco and a flame or hot wire lighter used to ignite it. Whether the flame be from a match or a lighter, this flame will inevitably start a fire if it comes in contact with suitable fuel. Thus, the discarding of a match into dry grass, before the match is extinguished, has been the cause of many exterior fires. Both match flames and lighters will cause explosions and fires with volatile flammables in the neighborhood. Many persons have attempted to commit suicide by turning on the gas, but then they wish a last cigarette. Sometimes it is the explosion that follows which is their executioner, but more often it puts them in the hospital with serious burns. It is important to realize that it is only the flame that causes this type of occurrence.

Cigarettes as a source of ignition of fires have been blamed in more instances than they should. A lighted cigarette will set fire occasionally to dry fuel on which it rests, such as upholstery, dry grass, and the like. Tests show that on top of most upholstery, a lighted cigarette will burn to the end without igniting it. A linear scorch will generally result but rarely a fire. If the cigarette is pushed partially between two upholstered cushions, enough heat may build up to cause ignition, but only under circumstances that conserve the heat without restricting unduly the air supply. Even on sponge rubber, which is very flammable, the gentle heat of the burning end of the cigarette continues to melt the rubber away from it, so that as the cigarette burns, it forms a trough in the rubber surface. Repeated attempts have not been successful in igniting the rubber. Thus, on the average, a great many lighted cigarettes will have been improperly discarded for each fire that results from such a discard.

Smoking in bed is an exception to the general rule that cigarettes pose a much lesser hazard than is commonly supposed. The reasons are simple. Bedding is often constructed primarily from cotton, which is readily ignited and is capable of sustaining a smouldering fire for an extended period. If the bed smoker should fall asleep so that a lighted cigarette drops into sheets, pillow cases, or other relatively flammable material, a fire can readily result, and statistics indicate that such fires are common. *Kapok*, sometimes used to stuff pillows, is especially susceptible to ignition—more so than is cotton. Feathers, on the other hand, are far less susceptible. Fires from this origin will rarely give serious difficulty in their interpretation because of the bed-centered origin.

A special hazard of fires that fall in this category arises occasionally when the sleeper is awakened by the heat but, being confused, adopts incorrect measures to control the situation. Such persons have been known to attempt to throw the burning bedding out of the windows or carry it to the exterior, and in the process have extended the fire over large areas of the building. If still small water could be used to control the burning until help can be summoned. If too large for these measures, the best course of action would be to close all windows and transom and leave the room, closing the door also, and to summon the fire department. Closing of the room serves the important function of restricting the air supply and secondly of localizing the smoke and heat. An open window and door, on the other hand, could set up a ventilation system that would rapidly spread the fire throughout the building.

An unpleasant accompaniment of bed fires is the danger of suffocation from carbon monoxide, with death or serious burns resulting as an aftermath. The fact that many bed smokers have also consumed unwise amounts of alcohol increases the total hazard greatly.

Another fallacy about cigarettes relates to the danger that they will kindle flame with volatile flammables. Repeated attempts to cause explosions by inserting a lighted cigarette into an explosive vapor-air mixture have resulted in failure even with the ash knocked off and puffing strongly. This is not proof that it cannot happen, but any claim that it has happened should be viewed with the greatest skepticism. The ash that surrounds the glowing end of a cigarette apparently acts in much the same manner as the screen around the flame of a miner's safety lamp. In any event, it either always or virtually always prevents ignition of such gas mixtures. This is not the case if a cigarette is lighted in such an atmosphere. In this situation the match or lighter flame is sure to initiate an explosion, fire, or both.

Pipe and cigar smoking are not basically different in their fire potential from cigarettes. The lighted pipe embers, like the lighted cigarette, will readily start a paper fire in a waste basket. It will also scorch upholstery or clothing, actually burning holes in cloth in many instances. It virtually never causes a flaming fire except with paper, packing, or similar material with the property of very ready ignition. The damage to clothing or upholstery is no less annoying when it is scorched or a hole burned through it but, at least, this will very rarely kindle the fire that burns the house. In any instance where smoking is claimed as the source of a fire, very careful attention must be given to assure that the claimed fuel is one that could be lighted by burning tobacco or, alternatively, the possibility of flame from a discarded burning match, which is actually a much more likely source.

ROLE OF ANIMALS IN IGNITING FIRES

The very mysterious origin of some fires may at times be clarified by considering the intervention of various animal species in carrying out unnoticed activities of their own. It has long been thought that rats may chew a match head and ignite the match. This explanation of fires from this source is apparently incorrect. Direct evidence has been found in various instances that the rat carries the match in its mouth and accidentally strikes the match head against some rough object such as a brick wall it is passing. The effect will be the same, naturally, since a fire can result from a lighted match, however it may be ignited.

Less consideration has been given to the possibility that rodents and birds may both carry lighted cigarettes, or other burning objects, possibly even to roof tops. Many rodents and birds have a strong instinct for acquiring and storing a variety of materials, either for nest building or other reasons. Paper objects are especially attractive, for example, to pack rats, whose nests have often been found to be lined with facial tissues. The paper cigarette may also fall in a similar category, although the author knows of no proven instance of such an occurrence. Some birds are known to carry sticks and other objects which they have been observed to drop down roof vents of heating appliances. Such a practice is clearly one that adds to fire hazard. Here again, if the object were a lighted cigarette and it was dropped on a flammable material, a fire could clearly result without intervention either of human intent or carelessness, other than the discarding of a cigarette which was not extinguished.

There appears to have been little or no study made of the habits of animal species in relation to the fire hazard, since such studies would be very difficult to carry out. Lacking more exact information, it is still apparent that fires must be started in this general manner at times, and more attention to this source of fire hazard is clearly indicated.

MISCELLANEOUS SOURCES OF IGNITION

Without nullifying, in any way, the generality that fires are invariably kindled by a small flame, electric spark, burning fragments, or other hot objects, there are some sources of ignition that do not fall neatly into these categories, or that are so unusual as to be somewhat controversial. In the former category are *lightning* and *spontaneous combustion*, and in the latter are *light bulbs*.

Lightning

Long before man learned to kindle a fire, it is almost certain that his knowledge of fire came from fires that were kindled by bolts of lightning. Lightning is a heavy electrical discharge of natural origin, which is fundamentally not different from the electric spark but is much greater in intensity. It is truly a result of a massive static charge, which also commonly causes small sparks that are hot enough to kindle explosions in explosive vapors. Such small static sparks would not produce enough heat to initiate an ordinary fire with, for example, solid fuel. When large enough, as is true of the lightning flash, the heat generated by the flow of electric current through the object struck will very often ignite it. This is especially true when that object is dry and readily ignited. The loss of buildings to lightning-induced fire is very large, although protective methods are available, and the loss has been markedly reduced within the last fifty or more years.

Trees are ignited only occasionally by lightning, the result being determined chiefly by their susceptibility to combustion. An old dead, dry trunk, for example, is much more likely to catch on fire than a living tree which is both a better conductor and less flammable. In general, many bolts of lightning will strike for every fire that is initiated. In regions where thunder storms are common, however, the risk is very great. In most instances, such fires are readily traced. When a fire of undetermined origin occurs in a remote area of mountain or forest, the investigation as to whether it was caused by lightning may be difficult. Meteorological records will generally establish whether or not there was a storm in the questioned area at the right time. Search of the area may uncover a tree or other object that was split or damaged by means other than fire itself. If this corresponds to the origin of the fire as indicated by study of the pattern, it is very probable that lightning was the origin.

The investigator needs to be familiar with the effects of lightning when it strikes objects. The most common effects are splitting of wood and mechanical disruption of structures. Most of these effects arise from sudden volatilization of water inside the tree or structure by heat from the electric current, so that the effect is that of an internal explosion. The effects are extremely variable, being determined both by the intensity of the electrical discharge and the localized condition of the object that is struck.

Spontaneous Combustion

Frequently, the claim is made that a fire has started from spontaneous combustion. This source would seem to violate the simple principle enunciated earlier and carries the connotation that in some mysterious manner fires just set themselves on occasion. The unmodified word "spontaneous" is at best very misleading and ambiguous, and it should be eliminated from the language as it applies to ignition of fires. The term "spontaneous chemical causation" would be far more accurate.

To the extent to which fires in this category actually occur, and this is certainly far less frequent than is generally believed, the most common origin lies in the fact that a chemical system subject to some exothermic reaction develops heat. Since the rate of the reaction generally doubles with each rise of about $10°C$., it will be seen that the reaction itself, by generating heat, tends to make itself go faster and faster with constantly increasing heat production. When there is sufficient insulation to prevent the rapid dissipation of this heat, the temperature may rise to a point of ignition. One of the common systems in which such effects are noted is the drying oil used in paint. Drying oils harden by oxidation at double bonds in the fatty acid chains of the oil, and heat is generated. For combustion to occur, the presence of considerable oil, spread in a film so as to have access to sufficient oxygen of air, and surrounded by enough insulation to allow the heat to accumulate, is required. Such a situation can occur in a large pile of "oily" rags, wet with drying oils, such as may accumulate during painting. It definitely will not occur with the hydrocarbon (lubricating) oils and is not expected with most fats and oils found in a household.

There are probably a considerable number of chemical reactions that in special circumstances, for example, cargo holds of boats carrying chemicals, are susceptible to effects such as this. If they catch fire as a result of an uncontrolled chemical reaction, this may be termed spontaneous because it was not caused by a human action; but the same can be said of many electrical fires, failures in machinery with generation of friction, etc., which are not termed "spontaneous" simply because their cause is readily understood by everyone. It would be equally appropriate to label lightning-caused fires as spontaneous.

Aside from occasional serious fires that occur in the storage of chemicals, the most common fires termed spontaneous are without doubt those that are ignited in hay, straw, or vegetable residues and manure piles. It is stated (3) that the annual loss from spontaneous combustion fires in the United States amounts to many millions of dollars. The same reference provides an excellent review of the knowledge regarding the course and mechanism of the effect. The phenomenon has been known since ancient times, and considerable experimental work has been performed in attempting to elucidate its mechanism (4,5,6). A case of spontaneous fire in a hay mow was observed and recorded in detail by Ranke who was himself a scientist and who performed further experiments in support of his theory of strong adsorption of oxygen by pyrophoric carbon. This was apparently the first detailed and well-documented description and served as the starting point for nearly all later work on the subject.

In order for vegetable materials to undergo spontaneous combustion, they must initially have a favorable water content, sufficient to allow fermentative processes to proceed. If the hay is well cured, it will not allow destructive fermentation, and if too wet, no fire will result also. Partially cured hay is favorable. The activity of the microorganisms will heat the hay to the thermal death point of the organisms as a limiting temperature. This is in the neighborhood of 160° F. and lower for many organisms. Regardless of insulation that exists in a large mass of the vegetable material, temperatures in this range fall far short of the necessary ignition point of the hay which is of the order of 540° F. or more.

This consideration makes clear the fact that the microorganisms raise the temperature to a point at which some exothermic chemical process can be initiated to raise the temperature much higher. It is this phase of the process that has been the subject of much attention and production of various theories, including the production of pyrophoric carbon, pyrophoric iron, heat from enzyme action, and even autooxidation of the oils contained in seeds. It is known that much acid is generated in the early stages of the process (cf. Firth and Stuckey), and that this is accompanied by marked browning of the hay. This high acidity has been used as a means of determining whether spontaneous combustion has occurred as contrasted with external ignition. Perhaps the most widely accepted theory is that of Browne who postulated the formation of unsaturated compounds by the action of microorganisms. These could form peroxides with oxygen of the air, later forming hydroxy compounds and atomic oxygen along with heat.

Whatever may be the actual mechanism, it is certain that it consists of a number of steps, the most important being chemical in nature. Other facts that are of greater importance to the investigator concern the surrounding observations that must be made. Spontaneous fires in stacked hay do not occur in less than ten to fourteen days after stacking, and generally requires five to ten weeks. The fire that occurs is in the center of the stack and burns to the exterior, usually forming some type of chimney to the exterior. A set fire in hay starts on the outside and burns toward the inside. Unburned hay in a stack in which spontaneous combustion has occurred will be very dark in color and have a higher acidity than normal hay. If a burning stack is spread apart, hay that is not on fire will suddenly burst into flame if spontaneous combustion is the cause of the fire. This is not expected to occur when the hay has been fired exteriorly. It must also be remembered that spontaneous combustion is an easy decision to reach when the cause of the fire is not known. It seems certain that many fires that have been attributed to spontaneous causes were actually set fires.

In addition to the two best known examples of systems subject to spontaneous ignition, oily rags and hay, a number of other fuel systems may ignite spontaneously. One of the more troublesome and common ones is coal, lying unmined in the ground or at times in large piles or storage bins above ground (7, 8, 9). As with hay, the moisture content has been found to be critical. Without doubt, the process itself is a slow exothermic chemical one in which heat is generated faster than it escapes in an environment which is well insulated. Another factor of importance is the type of coal, some grades of which ignite far more readily than others.

Esparto grass has shown to be subject to self-heating (10) under certain conditions, including proper and relatively high moisture content and high ambient humidity. Unsupported claims have been made for self-ignition of many other materials of vegetable origin, such as manufactured fiber board. It would appear that such claims would rest on the availability of a suitable environment for bacterial growth in the material, since this seems to be general with vegetable materials. Such a material as pressed fiber board would not appear to qualify if this is a proper limitation of the possibilities. Sawdust piles and other similar materials of wood origin have also been thought to ignite spontaneously, but there does not appear at present to be any convincing proof, nor are there any controlled studies on the subject.

Electric Light Bulbs

Probably the rarest of all potential sources of fire ignition, the electric light bulb, must not be overlooked since it is a hot object. Bulbs of low wattage normally operate at a temperature that will not burn the hand. As the wattage is s increased, and especially with bulbs in housings of some type that restrict ventilation, temperatures can build up to quite high values. Thus, potentially, a large light bulb can start a fire if it is in contact with suitable fuel. The reason that such events are very uncommon is not alone from the lack of a high enough temperature, but even more because something easily flammable must be in contact with the bulb, and this situation rarely occurs. It may happen in basements, storage rooms, and shops that packing material is allowed to be in contact with a bulb that is operating. The only fires known to this author, that are presumed to have had this origin, occurred under these conditions.

Experiments have shown that a bulb of 100 watts or more will cause discoloration and ultimate charring of cloth, paper, and similar easily affected material. At the same time, it was found nearly impossible, experimentally, actually to kindle a fire with bulbs of 150 watts or less. However, if the bulb can produce the results it does, it is a definite fire hazard when not kept away from readily flammable materials. In one instance, a large fire was almost certainly kindled by a 1000 watt light bulb which came into contact with nominal one-inch board flooring. The bulb was being used to keep a small compartment very dry, so that the wood was essentially devoid of moisture and probably more readily ignited as a result. The heat from such a large bulb is excessive and quite capable of kindling a fire. Such possibilities must be recognized and not underestimated by the investigator.

References

(1) McNaughton, G. C. *Ignition and Charring Temperatures of Wood*. U.S. Dept. of Agriculture, Forest Service, Nov. 1944.
(2) Graf, S. H. *Ignition Temperatures of Various Papers, Woods, and Fabrics*. Eng. Expt. Station, Oregon State College, Bull. 26, Mar. 1949.
(3) Browne, C. A. U. S. Dept. of Agriculture, Tech. Bull. 141, Sept. 1929.
(4) Ranke, H. *Liebigs Ann. Chem.*, 167, 361, 1873.
(5) Hoffman, E. J. *J. Agr. Research*, 61, 241, 1940.
(6) Firth, J. B. and Stuckey, R. E. *Soc. Chem. Ind.*, 64, 13, 1945; 65, 275, 1946.
(7) Hodges, D. J. "Spontaneous Combustion: The Influence of Moisture in Spontaneous Combustion of Coal." *Colleiry Guardian*, 207, 678, 1963.
(8) Berkowitz, N. "Heats of Wetting and Spontaneous Ignition of Coal." *Fuel*, 30, 94, 1951.
(9) Smirnova, A. V. and Shubnikov, A. K. Effect of Moisture on the Oxidative Processes of Coals. *Chem. Abs.*, 51, 18548, Nov. 25, 1951.
(10) Rothbaum, H. P. "Self-heating of Esparto Grass." *J. Appl. Chem.*, 14, 436, 1964.

10

Automobiles and Boat Fires

AUTOMOBILES

Although automobile fires are not uncommon, their incidence is actually small. This is surprising in view of the fact that the automobile constantly carries highly flammable fuel. The reasons that automobiles do not present a greater fire hazard are clear. Except for the gasoline, there is very little that is flammable in a motor vehicle. What there is, is essentially all concentrated inside the passenger compartment—the upholstery and interior finishing. This material differs little from the furniture in a living room in respect to its fire hazard. Cigarettes properly placed in the upholstery may start a fire, but generally only local damage will result from contact. The plastic materials commonly used in automobile upholstery tend to melt, but they burn poorly except when excessively heated. Foam rubber is perhaps the most hazardous material used in modern upholstery because it ignites rather easily and burns fiercely.

Any fire started in the passenger compartment will generally limit itself to that compartment. It is highly unlikely that the gasoline in the vehicle will ignite, and the remainder of the materials present are nearly immune since they are metals which burn only with enormous difficulty. Naturally, all of these statements must be modified by the presence of extraneous materials which may be flammable and are carried in the vehicle. Such fires can hardly be attributed to any quality of the vehicle other than that it is capable of carrying them.

Fuel Tanks

A tank containing gasoline would be considered an inherently hazardous object. The facts argue otherwise. If the cap is removed from the tank and a match is used to check the gasoline level, a fire and/or explosion may be expected, and properly so. Correctly capped, however, it is nearly impossible to ignite gasoline in a fuel tank. This is because the vapors of gasoline are much heavier than air and tend to fill the tank completely to the top. Also, except at low ambient temperatures, the vapor pressure of gasoline is such that it tends to form a mixture with air in the tank which is above the explosive limit and cannot be kindled. Strangely enough, this is more certain in warm than in cold weather in which this hazard is actually increased. It follows that fire will not ordinarily invade an open gasoline tank, but rather will be limited to the outside of the opening where there is sufficient air for combustion. With the cap in place, the small vent hole will not normally feed a fire, although it is not recommended that a match be held to it to find out.

The major hazard of the gasoline tank is its rupture in an accident. This makes available on the exterior all of the highly dangerous qualities of gasoline admixed with air and can readily lead to explosions and fire. The general placement of tanks in the rear of the vehicle and within at least some protective metal parts is responsible for the low incidence of gasoline fires from fuel in the tank.

A special case, however, that offers additional hazards is the *fuel tank of the large truck*. Most such vehicles are drawn by a tractor or truck-tractor, which must carry its fuel, and in a location that does not interfere with the normal function of carrying a load or pulling a trailer that carries the load. Such tanks are frequently installed on the sides of the vehicle, just back of the cab, and relatively exposed. In a sidewise impact, this tank is likely to be ruptured, as is true also if the vehicle overturns. Fire commonly follows such accidents that cause the release of considerable quantities of liquid fuel. Since most large trucks use diesel fuel (similar to kerosene) rather than gasoline, the hazard is somewhat reduced but far from eliminated.

Carburetors

The other portion of the fuel line where there is possible exposure to air and fire potential is the carburetor and its accessory attachments, such as the glass filter bowl through which, in some automobiles, gasoline passes to the carburetor. Here also, there might appear to be more hazard than is confirmed by the facts. A normal carburetor is sealed very well from admixture with air, except for the functional admixture without which the motor could not operate.

An under-hood fire in an automobile is almost certainly a gasoline fire, because gasoline is normally the only fuel, excepting a minor amount in the form of insulation of the wiring. Carburetor fires may be recognized by their pattern. In general, they tend to show heat effects in the upper portions of the hood compartment, and sometimes they are almost entirely localized on top of the motor. At other times, if an engine is operating and a leak in the neighborhood of the carburetor develops, the fuel pump will force gasoline through the leak in considerable quantity, and excess liquid gasoline may run to the ground or floor below the motor. This can ignite tires and other fuel in the neighborhood. However, a fire pattern that predominantly involves the *top of the motor* is almost certainly a gasoline fire.

Fires of this type are almost without exception due to broken fuel pipes, blown gaskets, or other defect which allows significant escape of liquid gasoline to the exterior, where it can be ignited by heat from the exhaust manifold, by sparks in the ignition system, or by a similar cause. In nearly every such incident, the mere presence of this type of fire serves as essential proof of a faulty fuel system. It remains to examine the parts to identify the item at fault.

To test this concept, a new carburetor was filled with gasoline, without an attachment to a fuel line. Efforts to ignite the gasoline in the carburetor were effective only at the fuel line attachment opening, and even here the flame was so trifling as to threaten self-extinction. Finally, to test the matter further, the entire carburetor, still filled with gasoline, was placed in a metal pan and liberally drenched with gasoline on the exterior. This was ignited so that the carburetor was surrounded with flame. The exterior gasoline burned away without detectable damage to the carburetor; the interior gasoline was heated enough to vaporize and contribute to the total fire through the open attachment for the gasoline line. However, when the exterior gasoline was consumed, the fire diminished to a small flame at the opening as before (see Appendix I). This author has seen melted carburetors, claimed to represent initial carburetor fires, but these have always been accompanied by exterior sources of intense heat, such as a burning garage in which the entire vehicle was totally destroyed.

Fires of definite fuel system origin have also been studied, and in no instance was the carburetor melted, although heat damage in the form of sooting and metal discoloration was very apparent, and gaskets were no longer functional. With the carburetor, as with other objects in contact with a liquid fuel-fed fire, only the portions exposed to flame are heated above the boiling point of the liquid fuel. At temperatures such as these, metals used in construction will not melt or even approach their melting points. When flames play on the metal without the cooling effect of contact with the liquid, the situation can be quite different.

The fact that only a defective carburetor or attachment will lead to a gasoline fire at that point has implications in actual fires which must always be of concern. For example, the glass filter bowl sometimes used ahead of the carburetor is more subject to breakage than other portions of the fuel system. When a crack is formed in it, gasoline can be pumped through the opening, flow over hot motor parts, and in fact form a layer on the floor under the car. When this deposit burns, it envelops the entire motor in heavy flames and produces great destruction. The significant fact, however, is that the system must have the defect (a broken bowl) before any significant hazard develops. Loose connections, e.g, on the fuel line, or any opening in the carburetor body itself or its gaskets, could cause this result. The situation is especially serious if the defect is one that does not lead to motor failure, because then the fuel pump can contribute relatively large quantities of leakage by the pressure it creates. If the motor is not running, the results are likely to be far less damaging, both because fuel is not pumped through a leak and because the gasoline that is lost is not likely to be ignited.

Electrical Origins

The electrical system of an automotive vehicle presents little danger of initiating fires. This follows from the fact that with the low voltages that are present, there is little danger of significant sparking. If a shortcircuit develops, the currents will be high enough to allow kindling of a fire, provided there is flammable or combustible material available at the point of heating. Such shorts will cause smoking or even flaming of insulation at times, and they do carry a danger. However, as pointed out above, the quantity of combustible material is small in an automobile and its distribution is not favorable for ignition from such a source. Thus, such fires are quite uncommon.

One additional source of fire hazard, occasionally present, is the battery. In use, it presents no fire hazard, but when receiving a quick charge, considerable volumes of explosive hydrogen gas are generated. An explosion will result from a spark or other source of ignition, and some fire may follow if the charging continues. Explosions, unaccompanied by fire, are the rule and even these are very rare.

Miscellaneous Origins

Perhaps most automobile fires are caused by carelessness in such things as leaving rags beside a motor, these rags containing oils and possibly gasoline; backing a car into a parking spot in which leaves or trash are afire; and many other similar thoughtless acts. Using a flame to search for a fuel leak or to check gasoline levels is another uncommon but very dangerous practice.

Arson is occasionally encountered, although a car is difficult to ignite. In some instances, considerable fuel is burned in the passenger compartment, which will completely destroy that portion of the vehicle but rarely affect any other portion. Insurance is generally so limited with automobiles that little motive exists from the standpoint of recovery of cash, but such instances do occur. Pranksters and hoodlums may deliberately ignite automobiles from motives of maliciousness or revenge. Here again, it is generally only the passenger compartment that is affected.

An additional potential fire hazard is associated with the common practice of carrying flares in a motor vehicle. Since these require ignition before they create any fire, their role in an automobile fire is primarily due to the additional fuel that they contribute, rather than to any inherent danger of starting a fire.

BOATS AND SHIPS

Fires in boats and ships are subject to exactly the same considerations as those in other comparable structures except for some modifications imposed by the difference in the environment. A large ship is like a large building of comparable degree of flammability, except for the fact that it is often loaded with large quantities of freight that impose their flammability characteristics on the ship. Such items as ammonium nitrate, used for chemical manufacture and fertilizer, are well recognized as a hazardous cargo, and special precautions are required. In bulk storage of flammables, smouldering fires are most common because of deficient ventilation.

Tankers, loaded with very large quantities of fuel, generally of petroleum origin, might be expected to be subject to enormously increased hazards. The main cargo of a tanker is no more dangerous than is the gasoline tank on an automobile and for the same reasons. However, the danger of spillage of fuel above board where the vapors can mix with air creates a similar hazard to spillage of such fuel in the exterior in any other environment. It does require that special precautions be taken. The other and far more serious hazard associated with the tanker is related to the possibility of rupture of the hull which then can release very great quantities of fuel. Under the conditions prevailing, it may be considered better that such escaped fuel be burned than that it remain on the water surface to spread over large areas, contaminate beaches, and create hazards to other shipping.

Because of the highly variable conditions that may exist on the ship, it is not feasible in this volume to attempt any extended discussion. The investigator with a good understanding of the principles of both fire and fuel and their chemical interpretation should have reasonable success in finding the correct solution.

Smaller craft, such as motor boats, are far more like automobiles than like fixed structures or even ships. The boat and the automobile both have a gasoline motor, fuel tanks of gasoline, and the accessories to the power system. There is one major difference that makes the boat fire more probable and much more difficult to investigate than the automobile fire. This is the fact that the flammables are all enclosed in a more or less tub-shaped container—the hull. Leakage of either liquid gasoline or its vapors will cause the fuel to settle on the bottom of the hull where it remains. Comparable leakage from the automobile is generally to a well-ventilated exterior. This is the basic reason for the greater danger and increased likelihood of explosion from fuel vapors associated with a boat. A secondary but important fact is that many boats are constructed of wood, while the automobile is not. A gasoline fire, for example, in a wooden container is a far more formidable phenomenon than the same fire around or even under a steel conveyance. The wooden hull is likely to be very largely consumed, at least to the water line, so that fire patterns are not distinct and points of origin are readily obscured. With larger wooden vessels which have considerable secondary interior structure, patterns will be found that are in every way comparable to those in any other similar structure. Such fires can be investigated in accordance with the principles that are detailed throughout this volume.

Supplemental References

Shifflet, G. A. "Investigating Automobile Fire Causes." *J. Crim. Law, Criminol. and Police Sci.*, 49, 276, 1958.

Davis, W. J. "Automobile Arson Investigation." *J. Crim. Law, Criminol. and Police Sci.*, 37, (1), 73, 1946.

11

Clothing and Fabric Fires

One other significant area of occurrence of fires is in clothing and fabrics. It is a truism to state that virtually all fabrics can be burned. However, the differences between the various types of fabric are so startling that the statement has little meaning. From the insurance standpoint, even a hole burned in table linen, upholstery, or bedding may be cause for a claim, but such an occurrence is so unlikely to initiate any larger fire that it is virtually of no consequence to the fire investigator. On the other hand, when a person's clothing catches fire with resultant severe injury to the wearer, the fire takes on another aspect. Such fires are sufficiently common to be a problem for both investigators and attorneys, as well as to the insurance interests. In fact, these fires have been the cause of passage of a number of laws regulating the manufacture and sale of fabrics for clothing. In these efforts to control the hazards of flammable clothing, almost all emphasis has been on the nature of the fabric and very little on the design of the clothing. That this is a fundamental error in many instances can be demonstrated readily.

TYPES OF CLOTH

Fabrics differ greatly in their properties with respect to combustion, ranging from some that are intrinsically very hazardous to others that cannot be burned at all. Those that are most dangerous are prohibited or controlled by law in some areas; those that are nonflammable are, in general, used only for special purposes that rarely include clothing. The types and their properties are outlined below.

Natural Fibers

Wool is employed in the manufacture of many fabrics utilized for clothing, as well as for bedding, upholstery, and miscellaneous purposes. Its fire hazard is minimal. It can be burned, but with considerable difficulty, and will not normally sustain a fire. Thus, wool blankets are used to wrap up persons whose clothing is on fire. About the only effect of exposing it to ignition is to char a hole in it but not to produce flames.

Other animal fibers and feathers share the general properties of wool since they are all of similar composition. Thus, they do not support fire; when burned in the flame of another material, they will char, decompose, and form combustion products with a characteristic unpleasant odor.

Cotton, although used extensively in the production of various cloths, many of which find their way into garments, is possibly the most flammable of all *common* fibers. Being akin to wood and paper in composition, and having a very large surface-to-volume ratio, i.e., a large surface for combustion, it is capable of being ignited with relative ease and will sustain a fire or allow it to increase greatly after being ignited. Its ease of burning is conditioned by several factors which will be discussed in some detail later.

Linen is being used much less than formerly, especially in clothing, and only occasionally is seriously involved in major fires. Its combustion properties are not greatly different from those of cotton, since it also is a cellulosic fiber, derived from stems rather than seed pods as is cotton. Because it is a bast (stem) fiber, it is much more variable than cotton, and this is expected to have some influence on its flammability. Cloth from predominantly fine fibers is expected to burn more freely than that from coarser fibers.

Kapok is not normally used as a cloth fiber but is involved in fire with sufficient frequency to make its fire hazards important. Like cotton, it too is a cellulosic fiber; but because of the structure of the kapok cell and its fineness, it exceeds cotton significantly in its flammability. It will ignite with ease and, when ventilation is available, it will burn with great intensity. It will also support smouldering fire very well. It is fortunate that its uses are limited so that it does not participate in clothing fires. It can, however, pose a significant hazard in the industrial areas, such as life preserver and pillow manufacture, where it is utilized.

Rayon (regenerated cellulose) is chemically very nearly the same as cotton, because it is made by dissolving cotton or other cellulosic material and regenerating the cellulose in the form of an extruded fiber by forcing the solution through a spinneret into the regenerating bath. Because of the similarity to cotton, the main difference in fire hazard is due to the different configuration and possibly size of the fibers. The rayon fiber is more or less round, whereas the cotton fiber appears as a twisted ribbon which has somewhat greater surface for combustion. Thus, in general, rayon may be expected to be a little less flammable than cotton, but not markedly so.

Acetate (acetylated cellulose) is similar to rayon in its manufacture and appearance, but the chemical modification of the cellulose diminishes its flammability. It melts rather easily but burns with some difficulty.

Synthetic Fibers

Synthetic fibers, which include the trade names Nylon, Dacron, Orlon, Saran, Estron, Dynel, Nylar, and Acrilan, can be treated pretty much as a group with respect to their behavior as fuel. All of them are synthetic compounds, but made with considerable differences in their chemical structures, so that in most respects their behavior is expected to differ. However, none of them is especially flammable, and nearly all of them tend to melt without flaming except in an applied flame. Thus, they tend not to support combustion independently and cannot be said to exhibit any danger as great as some of the natural fibers. Exceptions may exist, and certainly some fibers will be more flammable than others, as are the plastics from which they are derived. It is not possible without extensive testing to evaluate their comparative hazard. But it is believed that, in general, synthetic fibers represent a lesser fire hazard than do any natural or regenerated fibers.

It has been shown by Martin (1) that the above generalization is not applicable in all instances. A survey he made showed that in Switzerland a treated nylon is produced that is highly flammable and has led to one death from burning. The reason for the flammability is clearly related to the dyeing and finishing of the fiber rather than to the fact that it is nylon, which is one of the least hazardous of the synthetic fibers in use today. This finding led to new Swiss regulations for the clothing industry concerning the testing of cloth for flammability.

Cellulose nitrate fibers have been used in the past for clothing, but probably are not so used at the present time. The material from which they are manufactured is similar to lacquer base and to smokeless powder, and is intensely flammable. It falls within the group of fibrous materials that are presently not acceptable under the existing laws.

Glass fiber cloths are manufactured and used mostly for non-apparel purposes, although clothing has been made from them. Drapery is one of the common uses in which their obvious non-flammability and fire resistance are of importance.

Metallic fibers, either consisting of metal only or in combination with plastic cores or coatings, find their greatest application as accessory decorative threads in cloth. Their flammability is very low or nonexistent as a practical hazard, although such threads have occurred in clothing fires observed by the author.

Asbestos, a naturally fibrous mineral, has been used in special applications such as theater curtains because of its great fire resistance and absolute nonflammability. It is unlikely to occur in clothing or common household textile material.

Treated Textile Fabrics

Textile fabrics are sometimes treated so as to be more fire retardant than the natural cloth. These fabrics are likely to be employed in buildings for decorative or other purposes where some substantial fire hazard is present. It is less common to treat fabrics used in clothing, although some such treatment would be desirable where both fabric and design predispose to easy flammability. Test methods for such treatment are discussed below.

The Fire Danger

It will be noted that only a few of the fiber types discussed appear to present very serious danger with respect to flammability, at least of clothing, and that one of the most common, cotton, is one of the most hazardous. Clothing fires are reasonably frequent, and the results to the wearer are often disastrous, especially with children. The reasons for the special dangers must involve factors other than just the type of fiber, although this is unquestionably an important one. Consideration will show that the two chief additional hazards stem from details of the following manufacturing operations:

Spinning. Yarn or thread that is tightly twisted will burn with less intensity than loosely spun yarns. In the manufacture of some cloths, tightly twisted fibers compose the threads; in general, these are the cloths that are most tightly woven as well. The degree of tightness in spinning is a definite factor in determining the flammability of the fabric made from the yarn.

Weaving. The manner in which a cloth is woven is very important in determining the behavior when the cloth is ignited. This follows from the surface exposure of the fibers which is roughly proportional to the rate with which the cloth will burn. A loosely woven cloth, with spacing between the threads and fibers, will always burn much more violently than a tightly woven cloth of the same material. For example, cotton canvas burns relatively poorly while cotton gauze is consumed with great rapidity by fire, although both are constructed from the same type of fiber.

Garment design. The factor of design has received little attention as a factor determining fire hazard. It has apparently not been considered in drafting legislation relative to flammability of clothing. Such legislation is limited to the type of cloth which is tested only in small pieces. As with all fires, the amount of fuel, the surface-to-volume ratio which is related to ventilation or air access, and the presence of combustible cloth over possible points of ignition which would allow rapid increase of flame must be considered as important. Ruffles, especially, are a source of hazard in design because they are often quite voluminous, with considerable fuel and excellent ventilation. Furthermore, they are rather common on clothing of children, who already are more likely than adults to experience clothing fires. Pleats and other items of clothing design that increase the amount of cloth and especially the number of loose layers, which are most conducive to rapid spread of fire, will remain a source of hazard in this regard.

Ignition of clothing may originate in a variety of ways, most of which are identical with the manner of kindling any other type of fire. Both electric and liquid cigarette lighters in homes and automobiles, as well as pocket lighters, are some of the more dangerous items where children are concerned. Matches may lose a flaming head when struck, and burning tobacco may drop on a dress. One of the more common and dangerous sources of ignition is the heating appliance against which a flaring skirt may be pressed without the wearer's knowledge, and the burning candle may readily ignite the flaring sleeve at the dinner table. Again, the main item that predisposes to ready ignition is the design, rather than the type of cloth, so long as the cloth itself is reasonably flammable.

TESTING OF FABRICS

It is clear that any procedure for testing the flammability of fabrics is an empirical one in which comparison is more important than any absolute criteria, all of which must be somewhat arbitrary. For empirical testing on a comparative basis, almost any reasonable method which can be applied reproducibly to standard samples will be satisfactory. Since nearly all fabrics are to some degree flammable, it is of interest that the legislators who have drawn the codes in the

State of California at least have recognized, but not given sufficient consideration to, the very wide range of flammability which includes nearly all textile materials. Their definition is reasonable, but not specific:

"It has recently come to notice that of the various natural or synthetic fibers adapted and adaptable for use in the making of articles as herein defined, some are so inflammable as to constitute a dangerous risk of fire and hazard of injury to persons and property. . . . . It is not feasible by statute to prescribe more specific tests than those herein prescribed, for it would appear that none such have yet been fully developed."*

Although the equipment utilized for the tests under the administration of the State Fire Marshal is highly specialized and extremely difficult to duplicate because of the minutiae of the specifications, the testing requirements and procedure can be briefly described as follows:

Two types of test are run: one to determine the relative speed of burning or the speed with which flame travels either over the surface of the material or through material which is entirely consumed by flame; the other to determine the relative ignition-resisting qualities of an article when subject to radiated heat. Also, woven fabrics are classified into two groups in regard to their fire hazard: first, those that have a brushed, napped, or piled surface, not made of animal hairs, presenting a greater hazard than fabrics with a simple weave; and second, fine, sheer or lightweight fabrics such as voile, marquisette, chiffon, veiling, netting, and similar diaphanous materials. Simple woven fabrics of conventional types are not considered to be in a hazardous category. It is also recognized that some fabrics which are coated with plastic or made from plastic may be dangerously flammable. Tests are specified for different fabrics on the basis of the group into which it falls, these groups being as follows:

Group I articles. All articles having a brushed, napped, or piled surface.

Group II articles. All articles which have a smooth surface and have been impregnated with an inflammable sizing, filler, or coating, or which are made from a solid, pliable film.

Group III articles. All fine, sheer or lightweight fabrics such as voile, marquisette, chiffon, veiling, netting, and similar diaphanous materials.

---

*(State of California Health and Safety Code, 19810 (d)). The tests developed as a result of this code, and administered by the State Fire Marshal, are available as Title 19-Public Safety; California Administrative Code.

Group IV articles. Any other articles of wearing apparel that are not included in Groups I, II, or III.

For Group I articles, test No. 1 is specified. Group II and III articles are to be tested by test No. 1 and No. 2, but there are no definite specifications for Group IV articles.

Specimens for test are required to be not less than two by six inches and are to be dried in an oven at 221°F. for fifteen minutes and cooled in a dessicator over fresh calcium chloride for not less than five minutes. For test No. 1, they are to be quickly removed and placed on a specimen rack inclined at a 45° angle, on which the specimen is supported by horizontal wires having a diameter of not less than 0.009 nor more than 0.012 inches, and placed not less than one-quarter or more than one-half inch apart. They are ignited by a specimen ignitor which may be either a gas flame of C.P. butane or an electric element. At least five specimens are to be run and in no instance is the flame to travel through the cloth or over its surface at a rate in excess of five inches in six seconds. Failure of any sample to fall within the test limit requires repetition of the test with at least five additional samples. The average for all the ten samples is to be six seconds or more, and no individual specimen in this set is to have a time less than five seconds.

Test No. 2, designed to determine the susceptibility to flame from a radiant source, is somewhat different. Here, five specimens two inches square are prepared and dried as described and placed on the specimen rack over a prescribed arrangement of electric heating wires. A precise timer is attached to the wires, and a current of exactly 10.0 amperes is passed through the wires. The time necessary for the cloth to ignite under these conditions is determined. Here, the time limit for ignition is four seconds; failure of any specimen to surpass this time limit calls for additional tests on five more specimens. To be declared acceptable, the average of the entire ten samples is to be five seconds or more and no individual specimen of the second lot of five is to be below four seconds.

It is apparent that specified testing procedures as established by law in this instance at least are totally arbitrary, because human judgment must draw a line between what is considered safe or unsafe, even though there is no break in properties that will define such an arbitrary line. It is sufficient that cloth samples vary greatly in their susceptibility to ignition; and though the test methods are arbitrary and empirical, it is possible within this set of standards to prevent the sale of highly hazardous cloths. No attention or similar degree of thoroughness has yet been applied to factors of design, which may, as shown above, be even more critical to total hazard than merely those of fiber type and weave. The rate of buildup of flame due to the factors of available fuel and adequate access of air can certainly lead to serious hazards, even with cloth that passes all of the standardized tests. Conversely, even cloths that are intrinsically dangerous as determined by the empirical tests may be used in such a manner as to present no special hazard.

It should be noted that some of the prescribed tests provide for mounting the cloth samples at 45° from vertical. This was also prescribed in Switzerland until it was studied further by Martin (2). He noted that a vertical cloth showed, as would be expected, a greatly increased rate of flame spread, and that those persons who had been fatally burned by having their clothing on fire were invariably standing. His comparison of testing methods gave rise to new testing specifications in that country in which the cloth is tested in a vertical position. He suggests that a reexamination of testing methods and further research in this area are critical if clothing fires are to be brought under control. Special attention needs to be given to flame-proofing methods as the best ultimate solution to the problem.

TESTS FOR FIRE-RETARDANT PROPERTIES

In addition to the type of test described above for general fire hazard of textiles, other tests can be important both to industry and to the user of flammable cloth products. Except for economic and decorative reasons, it has never been logical for people to wrap themselves deliberately in layers of highly combustible material and to live in an environment in which fire is a constant component. For this and other reasons, industry has attempted to develop methods of coating or treating cloth to make it less flammable than would otherwise be the case. Considerable success has attended this effort, although most of the materials used tend to be removed by washing and dry cleaning, and the effort must still continue. Treated fabrics have been used most where special hazards exist, as in decorative fabrics on the interior of buildings, and where they are exposed to a minimum of cleaning operations. In order to test the flame retardant properties of such fabrics, the American Society of Testing Materials (A.S.T.M.) has set up testing specifications and methods which will be briefly described as follows:

Samples of not less than two square yards of fabric are selected; test specimens two inches wide and twelve and one-half inches long are cut from various parts of the sample. Ten specimens are tested, each being suspended by a clamp and placed behind a vertical sliding glass front. A luminous gas flame from a laboratory burner with a three-eighth-inch tube, adjusted to a height of one and one-half inches, is placed below the specimen which is suspended vertically three-fourths of an inch above the burner top. It is applied for twelve seconds, and is then withdrawn. The length of char from this exposure is measured, and the comparative flame retardation is inversely proportional to the length of the char.*

---

*The details of the test method are available as A.S.T.M. Designation: D 626-41 T, issued in 1941.

Other approaches to the classification of textile flammability have been offered. Perhaps that of Segal (3) is as useful as any. He sets up five general classes, described in "broad descriptive terms" which are still useful. These are: (1) extremely flammable, (2) highly flammable, (3) moderately flammable, (4) slow-burning, and (5) flame resistant. He stresses that the physical form and construction of the fabric are most basic in determining the flammability, as well as the thermal properties which are of greatest significance with synthetic materials.

In the extremely flammable category, he emphasizes cellulose or rayon fibers woven into a physical form most conducive to rapid burning, such as very sheer, lightweight, or open-textured material or one having a deep-brushed, napped, or piled surface exposing countless tiny fiber ends. He also includes some very lightweight, sheer silk fabrics such as chiffon.

The group of highly flammable fabrics has a slightly reduced degree of flammability, established only by standard test methods. Those fabrics that barely pass the tests would be classified in this category, but qualified by the statement: "It is questionable, however, whether the distinction would have much significance for the cloth-fire."

Flame-resistant cloth would include any cloth that when exposed to a direct flame shows no flame spread past the point of exposure. These would include pure wool and a few synthetic fibers. Slow-burning cloths would consist chiefly of wools so woven as to allow them to support fire to a limited degree or heavy, flat types of cotton, both of which ignite with difficulty and burn slowly. All other cloths would be classified as moderately flammable. It is noted that most cloth in use falls in this category, and that almost all fabrics involved in fire belong in it also.

One important hazard of most synthetic fabrics not covered in the classifications discussed is their tendency to melt when heated, often with burning at the surface of the molten material. When melted, the drops tend to cling tightly to the skin of the wearer so that they cannot be readily detached. Very severe burns can result from this effect, even with fabrics that are classified in the slow-burning categories.

References

1) Martin, E. P. "Zur Frage Der Brennbarkeit Moderner Textilgewebe." *Chimia*, 18 , 48-56, 1964.
2) Martin, E. P. "Zur Problematik der Kleiderbrande." *Chimia*, 22, 195-201, 1968.
3) Segal, L. "Classifying Textile Flammability." *Fire J.*, 60, 20, 1966.

12

Practical Investigation of Structural Fires

Before the investigator attempts the actual investigation of a fire, he should be reasonably familiar with the principles of fuels and of fire behavior outlined earlier. In addition, he must have a clear idea of the purposes and goals of the investigation. Then he is ready for the routine of the various examinations that must be carried out.

The *purposes* of the investigation, in logical sequence, are as follows:

1. To determine the origin of the fire in space. Where did it start?
2. To determine what was the causative agent, i.e., the nature of the initial fuel and the nature of the ignition. What ignited it? What was ignited?

3. Was it an intentionally set fire or an accidentally set fire? Generally this question is answered when the answer to No. 2 is clear.
4. If it was intentionally set, was it an act of arson or was it normal to the surrounding circumstances? Since this question involves legal concepts, some consideration of this point may be necessary from the attorney. A man may burn his own property under certain circumstances with no intent to defraud anyone, but merely because he wishes the property destroyed. In this respect arson, more than most other crimes, requires that the investigator establish the *corpus delicti*, which in many crimes is immediately apparent, and then prove the case as well.

The time element is always important, and often critical, It may be summarized as follows:

During the fire. Though never essential, it is often helpful for the investigator to be present if possible during the fire. The exact region in which it starts may be apparent, but ordinarily is obscured by smoke and heat which prevents easy entry into a burning building. In the course of fighting a fire, it is uncommon for the firemen who are physically present to do more than generally estimate the region of the building in which the fire originated, because the fire itself produces so much smoke and lethal gas that the necessary close approach to the local origin is impossible. An investigator on the premises at this time suffers from the same difficulties, but he can profit by observing the activities of the firemen which generally alter the course of a fire as well as its intensity. The investigator can also photograph the sequence of the fire, although others generally do this anyhow (see section on "Photography of Fires and Fire Scenes").

Immediately after the fire is extinguished. So that access to the remains is available, it is always well for the investigator to be on the scene at this time. While the condition of the remains is often not suitable for anything approaching a complete investigation, it may then be possible to use a hydrocarbon detector (see section on "Detection of Flammable Liquids") and obtain essential proof of arson if such materials are present. It is also possible at this stage to obtain samples which may contain such materials for later laboratory examination. The general consequences of the fire may be more apparent at this time than will be true later when the rubble is cleared out. In some fires, it is virtually impossible to make any examination until some clearing is accomplished because of the large amount of collapsed rubble present. In small fires, however, much can be accomplished at this stage because when the fire is localized, so also is the damage, and reasonable access is available. In the largest fires, it may be days before a truly complete investigation may even be initiated, much less finished.

During clearing of the fire scene. It is well for the investigator to be on hand. Not only will he assume some supervision of the clearing operation, but his presence at this time will make him aware of anything significant that may be uncovered. Disturbance and loss of significant items of evidence frequently occur in clearing a fire scene, and they will be avoided only when the investigator is present and able to take the necessary steps to preserve such items. Unfortunately, it often is not possible for the investigator to spend the time necessary for this important duty, and coordination between the people who are concerned with the clearing and those concerned with the investigation is not always of the best. With all due allowances for the difficulties, it is highly recommended that the investigator be present at this stage whenever circumstances allow.

After clearing of the fire scene. The investigator at this time has a clear chance to see all significant burns, provided only that some have not been removed in demolition that often accompanies the clearing operation. For example, while dangerous walls may have to be removed and superstructure demolished, the bulldozer that is probably used may also tear out bottoms of walls where burns exist to floor level. Fortunately, destruction of the most significant portions of the pattern by demolition is rare as long as foundations and bottom structures that are still firm are allowed to stand. In many investigations, this is all that is available, especially as an accompaniment to civil litigation. The investigator must not be less thorough because much of the building is removed. Nor must he overlook any remaining rubble or specific items removed from the structure and retained. It is often possible to recover highly significant evidence from such material.

Investigation after extensive demolition or repair is necessarily handicapped to a considerable degree. Often there are photographs available that will assist in evaluating the limited remains of the fire scene. These can be very helpful, but they do not substitute for original investigation at the scene. In a few instances, it is not possible to arrive at firm conclusions as to the cause of a fire under these circumstances; more often the astute investigator will be successful, even with this increased difficulty. Only rarely can he hope to detect the use of liquid flammables except from the pattern of the fire, but this is sometimes so distinctive as to leave no doubt as to whether or not they were used. Many investigators working under these circumstances are tempted to limit their activities to questioning of witnesses. Despite the occasional gems of information that are acquired by this means, most of the questioning will not yield definite and reliable information (see section on "Interrogation of Witnesses").

It is evident that *the most productive investigation will be done at the scene of the fire*, regardless of the time sequence of that investigation as related to the fire. It is also clear that no invariable routine can be laid down for the reason that there are too many variations in the circumstances, both of the fire and of the investigator's relation to it. Nevertheless, with whatever situation the investigator is confronted, there are certain things he will need to ascertain or evaluate, and there is an orderly method of approaching the investigation. Such a general course of action is outlined below, with the understanding that it will require modification to suit the conditions and circumstances that prevail. There are also a number of items that will be required for the investigation and that should be routinely available. These are as follows:

1. Camera equipment, as outlined elsewhere, including not only the camera itself but, if possible, electronic or other type of flash equipment and slave lights.
2. Portable illumination. Because there will rarely be electricity in a burned building, it is necessary that battery-powered equipment be available for inspection. Strong flashlights are useful, but a more general source of illumination such as the Sun Gun is desirable. These units have short battery life, so they must be turned on for short and critical use only, relying on longer-lived units, such as flashlights or battery lanterns, for general exploration. Gasoline lanterns may be employed but can also create an undesirable hazard of starting a second fire in the frequently flammable debris left by most fires. In very large fires it is sometimes possible to obtain a motor generator unit that will supply regular electric current for conventional illumination.
3. Tools for clearing debris, which may include a shovel, rake, and broom, along with pliers, tin snips, and saw for removal of wires, wooden pieces, and sheet metal.
4. Measuring equipment, such as a one hundred-foot steel tape and possibly a shorter one of eight or ten feet.
5. Record-keeping equipment, which must include a notebook but may also include a portable tape recorder. Many more details can be readily recorded with the tape recorder than in a notebook, but the latter is essential for making diagrams.
6. A hydrocarbon detector, especially for very recent fires.
7. Sample collection equipment. Included here are jars or other containers that can be sealed in case samples of soil or debris are to be tested for liquid accelerants. In addition, there may be various items that should be preserved for more detailed examination. Paper bags, cartons, or similar containers are suitable for the larger items, cellophane or plastic bags for the smaller.

138   *Fire Investigation*

DETECTION OF ARSON

Only occasionally will the exact cause of a fire be known prior to its investigation; in this fortunate circumstance, it is likely that no investigation is required. The large group of fires of undetermined origin will include many that have been deliberately set, so that investigation to uncover arson is a normal and necessary portion of every fire investigation. Aside from some special considerations that are covered in the succeeding chapter and which relate specifically to arson, the principles laid down for general fire investigation do not differ in any material respect from those applicable to investigation of arson alone. It is especially important, as outlined subsequently, that the investigator be familiar with the limitations imposed on the arsonist and with the detection of the steps taken by the arsonist to overcome them. In all other respects, the investigation of a fire is the same as the investigation of a suspected arson.

It is also necessary that the general principles of combustion and the properties of fuels be understood, especially in order to make the distinction between the accidental fire, which may have some features characteristic of arson, and the arson that has features similar to those of the accidental fire. Thus, negative as well as positive indications in the fire must be considered equally. Every aspect of the investigation that bears on this important point must be uncovered and considered.

GENERAL ROUTINE

A routine of investigation that is generally applicable or subject to suitable modification is as follows:

1. The burned structure is to be viewed first from the outside, preferably from all sides, and evaluated as to the following points:
    (a) Extent and intensity of the fire as indicated by general destruction.
    (b) Character of the destroyed portions such as roof and walls.
    (c) Outlets of fire from a structure, e.g., through windows or hole in roof.
    (d) Type of exterior construction, roof, walls, and other residual structures.
2. A rapid but general survey is made of all interior accessible burned regions to obtain an over-all picture of the development of the fire and the involvement of various portions of the structure. This information is for background primarily, and it is of limited utility in determining the nature or origin of the fire.
3. Direction of spread of the fire will be noted during this survey. It will be upward, partially lateral, rarely downward, but its direction will indicate the general region of origin when properly interpreted. This should be, and generally is, close to the low point of the burn.

4. A specific search is made for the low point of the fire. In many instances, fire has burned from or close to a floor which is intact, and there will be low regions rather than a low point. A suitable routine for establishing a probable point of origin may be outlined as follows:
(a) The lower region of the fire, such as the lowest story of a multistory building or the lowest floor underlying the burned area, is located by carefully examining the entire burned area. Rubble will be removed as necessary to uncover low burn. A burn covered with rubble is highly significant because it occurred before the fire was well developed, and it was quenched by the falling of fire residue from above.
(b) All low points will be recorded and photographed. Sometimes several such points and sometimes entire regions show burn from the floor upward.
(c) Low points are then evaluated with respect to the total pattern. One or more is likely to be the origin, but it is not always simple to determine which it is. Special attention is given to whether or not the burn extends exactly to a non-burned floor, or starts a little above it. If liquids were used and reached the bottom of a burned or charred wall, the burn will be exactly to floor level. In all other instances, it is unlikely that such a burn will appear.
(d) Single low burns, especially when they are quite deep, will almost invariably indicate that this is the origin, and the remainder of the pattern will normally be consistent with this origin. When there are multiple low burns, careful evaluation of the appearance of each, and of its relation to the total pattern, is essential. Multiple low burns may also indicate multiple origin, characteristic of deliberately set fires.
(e) Low point(s) of burning must be carefully checked to eliminate the possibility that the burn was actually from above and due to falling material that was either very hot or blazing. Floors may burn through from above because of a collapse of a burning roof, for example, and this may be the low point of the entire pattern. The key to the determination generally is the inspection of the upper and lower burned floor, etc. If the burn was from above (blazing material), it will have charred and penetrated partially all around the top surface of the penetration. If from below, it was a true point of origin. Another check is the occurrence of gaps in structures above, where they have fallen. A floor burn directly below a gap in the roof or floor above is immediately suspect as having been burned from falling material.
5. Having located the probable origin, a search is started for an initial fuel. As indicated previously, a hole burned *upward* through a floor above the ground is virtual proof that solvents penetrated the floor through an opening and burned below the floor. If the low point is at floor level, a

penetrating burn at the junction of the wall carries the same inference. Other low points generally do not indicate the use of liquid flammables, although char patterns or discoloration on the floor may at times imply such use. Often, the initial fuel is merely structural material or trash burning against it. The appearance here is quite different from fires started with accelerants. Electrical fires ordinarily fall into this category.
6. Samples are collected, or a hydrocarbon detector is used at points that indicate the probability of use of accelerants.
7. Source of ignition is then searched for. With accelerants, it is not likely that a source will be found, because it is probably a match which has been destroyed and its remnants lost. Sometimes, however, even this source of ignition is found. Special ignition devices, if present, will ordinarily be located readily. However, such devices will not be present often.
8. Electrical origin will be indicated by the type of local burning, immediately adjacent to electrical equipment or wiring and generally quite deep. Such burns may be difficult to find because they occur inside areas that are not opened for inspection. In one instance, the indications were under an unburned rug where the floor and the rug pad were both charred in exactly the same pattern as an extension cord under the rug. This was not the actual origin, but it indicated the direction in which it was found.
9. Residues and rubble that are moved should be inspected carefully. Fallen timbers will show the pattern of the fire that burned them; when the timbers are visualized in their original position, these patterns may contribute greatly to the knowledge of the fire. Such timbers may be sawed or cut loose and saved, or at least be well photographed. Among rubble will often be found items of furniture and secondary installations from the structure. These may be of great value in analysis of the fire pattern. Significant items of this type need to be retained or at least be well photographed.
10. Containers are commonly found at fire scenes. While most of them will be normal contents of the house, such as paint cans and household supplies, they may have contained flammables which could be associated with arson, and if not, may still provide useful information. If the can contained a liquid, even in small amount but up to a full can, and it was tightly sealed and heated, it will be bulged and split around one or more seams. If the liquid was flammable, it has boiled out of the can and contributed substantially to the total fuel. If such a can is not bulged or burst, it is indication that it was either totally empty or that the cap was loose. Under some circumstances, lightly soldered cans may selectively heat from radiant fire above them, so that the solder softens and allows escape of the content without showing a bulge. In a suspected arson, it is

highly essential that all such containers be collected and studied with the greatest care.

11. When multiple origins are present and arson is strongly suspected, it is quite essential that these points be rechecked with respect to original access to the parts of the building that contained them. For example, one origin may be in a store to which the proprietor has access. This could throw suspicion on him as an arsonist. If another origin is present in another place in the building to which this individual did not have access, serious doubt would immediately be cast on such a suspicion.

12. A final portion of the general investigation of the fire scene involves a resurvey of the entire site in view of the presumed finding of point of origin, source of fuel, and possibly ignition. This is an important portion of the investigation because, if there are discrepancies in the theory that is now developed, they can be found and reconciled or the theory modified. Such matters as these can be used by the opposing attorney in any subsequent trial that may occur; failure to resolve all the contradictions can be highly embarrassing.

During this final phase of the general investigation, it is wise to note specifically any types of building material which may have enhanced the development of the fire and to make certain that one clearly understands the plan of the structure. Although these details may have all been noted in the earlier investigation, failure to fix them in the memory or the notebook can later raise problems that are better avoided. It is also possible that this review of the inspection will turn up unnoted details that should be known, although they have little bearing on the theory or interpretation of the total fire.

SPECIAL ITEMS OF INTEREST

During the performance of the investigational routine described, it is clear that the investigator is constantly noting indications of: (1) fire pattern, (2) extraneous fuel, and (3) sources of ignition. Inasmuch as electrical failure is a frequently claimed source of ignition, some special consideration of this matter is indicated.

Electrical Wiring

To check out an electrical system in a burned structure is often the most frustrating portion of the entire examination. Wires, conduit, and remains of fixtures are often scattered throughout much of the bulk of rubble. Wires will generally lack insulation, this having been burned off in the fire. Since wiring is most commonly of copper, and copper when heated is oxidized strongly to copper oxide, the blackened and brittle condition of the wires will make them difficult to handle when attempting to trace them. The most significant item to look for in such an instance is not just a lot of damaged wire but rather fused copper in localized areas. These are most likely to occur in the neighborhood of switches, outlets and similar connections, and at the fuse box. At any of these places it will be common to find many wires with rounded, fused ends, demonstrating that there was overheating, almost certainly by shortcircuiting. When there are numerous such fused ends, it is unlikely that a fire origin has been found but rather that there was failure of the breakers to shut off current on shorted wires farther away from the breaker or fuse box. When there is a single heavy fusion, but other portions of the wiring are generally intact, this single point becomes highly suspect as a source of the fire. Normally, in such an instance, the fuse is blown or the breaker open, indicating that this shortcircuit occurred and initiated the operation of the safety mechanism, but not before a fire was started.

It should be noted also that wire within a metal conduit will rarely short except when the conduit is heated sufficiently by an external fire to burn away the insulation. If a shortcircuit is formed within a conduit from any cause, it must be a very heavy one to penetrate to the exterior and start a fire, although such events are not unknown. To do so requires that the conduit itself be heated to the point of fusion. In a very rare instance, it might be heated above the ignition temperature of some material in contact with it, which could become ignited. The high heat conductivity of the conduit itself gives considerable protection against this type of event. Thus, a conduit is actually a very good protection against starting of fires from wires enclosed in the conduit.

One additional effect that has been noted in more than one instance involves the zinc used in galvanizing iron conduit. Zinc has a very low melting point and will melt in a fire surrounding the conduit. The heat of the fire may also remove the insulation, so that molten zinc will contact copper. When this occurs, the two metals will form brass. Either the molten zinc or the brass formed may complete a shortcircuit between wires that were not previously in contact. Such an effect is not a cause of the fire but the result of an external fire, and it should not be interpreted otherwise. The presence of brass within a conduit or spilled from a broken conduit is actually very good evidence of an external fire which led to its formation.

Appliance Condition

The condition of an *appliance* that has started a fire is not expected to be identical with one that has been in an exterior fire which it did not start. The exterior fire will damage any appliance by burning the enamel, discoloring and corroding the steel, and otherwise distorting or warping it. If the fire is exterior to the appliance, whatever damage it suffers will show a uniformity, or at least an agreement with the fire distribution in the vicinity as shown by adjacent structures. If the appliance started the fire, it will have some local damage or effect that is quite distinct from this general damage pattern. This will be a point of consequence to the investigator.

AREA PROTECTION

In any fire, the greatest damage possible will be at points where active flame existed for an appreciable time. Remembering that flames always burn upward, it follows that floor areas will rarely receive much active flame. They will char by radiant heat in most instances, but it is often noted that areas of the floor are totally undamaged though surrounded by damaged area. Such areas tend to have a definite form—circular, rectangular, etc.,—and invariably show clearly the location of an object sitting on the floor at the time of the fire. The object may be flammable and show the effect as well as though it were nonflammable. Cardboard cartons often serve to protect the floor under them because the bottom of the carton does not burn well due to poor ventilation, even when the remainder of the carton is totally destroyed.

The importance of such protected areas lies in two facts: (1) locations of the objects are definitely determined, and (2) the general type of the object is defined. For example, a bucket that originally contained liquid accelerant might be placed on the floor in the fire area, and its presence would be proved. Delineation is generally so accurate that the size and shape of the bottom of the object is readily determined and compared with any suspect item that is located. Naturally, the objects are frequently removed during the fighting of the fire or in clearing the area afterward, and thus will only occasionally be found still sitting in the original position.

In some fires there is a question as to whether or not such items as flammable adhesives, which were being used prior to the fire, may have led to the ignition. This type of problem will be clarified greatly by establishing that the container was actually present and exactly where it was. The area that protected the floor may, in such an instance, be very critical to the investigation. For example, if the container was present and open, it is certain that it would lead to an intense local fire immediately above it. If the floor area is in agreement with the questioned container as to size of the bottom, and there is an area of intense burn above this site, it stands as strong evidence for this set of circumstances. Lacking any of the noted elements, it is equally convincing against this reconstruction.

Another situation in which a protected area is important is in the claimed *waste basket* or *trash box fire*. When there is reason to believe that a fire may have been initiated in waste material in such a container, it is virtually certain that the bottom of the container will protect the area under it, and its position may then be accurately determined if the floor is discolored or charred around it. By examining the arrangement adjacent to the trash container and the extent of fire in this region, it may be possible to assign the cause of the fire to this origin or to rule it out of consideration. Without the knowledge of the exact location of the container, this is likely to be impossible or very difficult. In numerous instances, a fan-shaped char pattern will occur on a wall just over such a point, with its lower boundary at the height of such a trash container. This circumstance is frequently the very best evidence of the trash container as the origin of the blaze.

Protected areas on walls are also frequently observed for similar reasons. The location of an item of furniture which is no longer present can often be determined accurately from such an area. The utility of such areas on walls is of lesser importance than on floors but should not be overlooked. Its opposite effect is also observed occasionally in instances where the furniture burned and collapsed; instead of protection of the wall behind it, there was enhancement of the burn. This effect will not necessarily serve to locate the position of a specific piece of furniture because the marginal dimensions are lost, but it can at times contribute significant knowledge of the fire origin, when considered along with the general fire pattern.

A combination of the two effects is also seen at times. When a piece of furniture is afire, it may both protect the wall behind it and, by intensifying the local fire, increase the amount of wall damage on the sides and above the item. This is especially important when the fire actually started in the furniture in question.

REMAINING RUBBLE

In clearing out the fire scene, there will nearly always be piles of partly burned rubble close by. Sometimes these are raked together, sometimes they are pushed together by bulldozer, and occasionally they represent the materials dragged from the burning building by firemen or removed manually at a later time. Such piles should always be searched with some care, looking especially for items that are significant and missing from the previously inspected scene of fire.

Commonly found in the remains will be partially burned furniture. Inasmuch as many fires start on or in the immediate neighborhood of furniture, close examination of such remains is often most informative. It has been known to alter entirely the preconceived idea of the fire origin as shown by the cleared area already examined. For example, in one instance, it was clear that a fire had started in the vicinity of a bureau, but was believed to have been in front of the bureau. When the bureau residues were examined, it was discovered that the fire had in fact started *inside the bureau*, in the drawer above the bottom one. One drawer was found to be essentially undamaged; one drawer nearly totally destroyed; another somewhat less, with the fire from below and in front; and a fourth still less, with the bottom intact. It was obvious that the bottom drawer had been below the fire, which had started in the somewhat damaged drawer and spread upward with increasing destruction to the top of the bureau. This pattern could also be confirmed by examination of the exterior portions of the bureau which were found separated and could be reassembled.

Another significant item of furniture is the bed, since many fires start in bedding. A smouldering mattress will nearly always be found in the rubble in such an instance, and the degree and character of the burn may be assessed. The same considerations will hold with upholstered sofas and stuffed chairs. When their portions can be recognized, the pattern of the fire that burned them will be very clear and will add immeasurably to the total information.

Containers both of trash and of liquids will often be located in fire residues. Here may be found the container that brought the accelerant or the trash container that was accidentially set afire. Paint cans may be present, and their open or sealed condition will indicate whether or not they made any contribution to the fire. Special ignition devices, if present, may be located in such piles of burned material. The possibilities are too many to do more than indicate some of the common ones. In any event, some of the most useful information may be lost if a careful search of debris removed from the fire scene is not made.

METALS

As indicators of the intensity of a fire, metals are often very valuable. While examining the fire scene, previously molten aluminum residues are often found. Since this metal melts at a comparatively low temperature, 1220°F., this is not unexpected or necessarily valuable when found. A more useful indicator is molten brass or copper. Depending on its composition, brass melts over a range of 1625° to 1780°F. Its presence in a burning structure is not unusual, but its melting temperature is sufficiently variable to limit somewhat its value. Copper, commonly present as water piping and electrical wiring, will only occasionally melt in any but the hottest fires, its melting point being 1981°F. To find molten copper in a residential fire (excepting from direct shortcircuits) is so unusual that when found it proves an exceptionally hot fire. In larger structures, and especially where there is much chemical fuel as in some industrial buildings, it is not uncommon to find previously molten copper that did not result from short-circuits in wiring. These fires are capable of generating much more heat than simple wood fires, however large. Whenever any residues of molten metal are present at the fire scene, they will reliably establish a minimum temperature for the point of their fusion in the fire. The investigator may use this fact to advantage in many instances, because of the differences in effective temperature between simple wood fires and those in which extraneous fuel, such as accelerant, is present.

Copper especially and copper alloys to a lesser extent are very subject to oxidation when heated in air. This results in formation of the black copper oxide on the surface of the copper object; if the heat and air exposure are continued long enough, ultimately the entire object, e.g., wire, may be changed to oxide. The degree of oxidation of exposed copper wiring can serve as a handy indicator of the conditions existing at such a point of exposure. Copper wire carried in conduit is not so subject to oxidation because of limited access of air and the general reducing atmosphere that results in the conduit from decomposition of insulating materials.

DETECTION OF FLAMMABLE LIQUIDS

It is customary and also good practice for firemen and investigators attached to fire departments to search for flammable liquids at each fire in which there is any reason whatever to suspect that arson may have occurred and even in accidental fires. The primary test is to smell all suspicious areas, since many flammable liquids have odors which are detectable and often characteristic. The opportunity to detect such materials by their odor is limited in time, because all that can be smelled is vapor arising from liquid, and vapors are rapidly dissipated. Thus, the only very effective time to smell such odors is immediately following the fire.

Another technique is to scoop up any soil or other suspicious residues which may contain flammable liquids and place them in water. Many flammable liquids are not soluble and will float on the water surface as a very thin film which will show interference colors. Such a film on the water's surface is exceptionally sound ground for the suspicion that flammable liquids are present, and it calls for further investigation and especially for collection of samples of the suspicious materials for laboratory examination.

Samples for later analysis should be recovered at the earliest possible moment. It should be remembered that the surface of soil or of absorbent debris is likely to contain no liquid, because any on the surface has been burned. By digging down below the surface, a sample may be collected that contains liquids, if any were present. This material must be placed in tightly sealed containers, such as the glass jars mentioned earlier, to prevent further evaporation of the liquid.

A *vapor detector*, if available, should be employed immediately after entry can be gained to the fire scene. This detector will also locate positions from which samples may be collected for laboratory examination. Several types of detectors are available commercially. Some are termed *hydrocarbon detectors*, although they may not be limited in sensitivity to hydrocarbons alone. A variety of operating principles may be utilized, depending on the type of detector available. These are not of direct concern to the investigator and will not be reviewed, since effective operation is his major interest. Operating instructions are delivered with all commercially available detectors and these should be followed exactly. It should be realized that detectors of this type operate by picking up fumes from the flammable liquid, and they will be progressively less effective as the volatility of the fuel is lowered. Thus, kerosene would be more difficult to detect than gasoline which is more volatile. In addition, volatile components of all mixed fuels tend to escape both during and after the fire at a faster rate than the less volatile components. This is a major reason why tests should be made at the earliest opportunity. It will also be helpful to scrape aside top layers of suspect soil or trash and to employ the detector with freshly exposed surface. A positive test should always be taken as sufficient reason for laboratory examination of samples from the area in which the tests are strongest.

PHOTOGRAPHY OF FIRES AND FIRE SCENES

No discussion of the photographic process is included here; rather, this section may serve as a guide to what should be delineated in the photographs, and how it should be done in different sets of conditions. The camera is virtually indispensable in a proper fire investigation, as it is in all other types of investigation of physical evidence. Thus, the investigator is most strongly urged to provide himself with adequate photographic equipment and facilities and to use them, whether he does it himself or hires someone to do it. In either event, *the camera should be constantly available.*

During the Fire

Two types of photograph taken during the progress of a fire are valuable: (1) general views showing as much as possible of the scene and from different vantage points, and (2) close-up views of the earliest portion of the fire that is available. The latter photographs will be difficult to obtain and often impossible because of smoke, heat, and other physical obstacles in taking pitcures. Thus, in many instances, only general views are possible, but *these should be made, especially during the early parts of the fire rather than in the later stages.* Press photographs are often available, but nearly always these are confined to the height of the burning when they appear most spectacular and indicate very little of use in determining the critical early stages. In addition, there is a tendency to take all photographs from a single good vantage point, which is a serious mistake. No information relative to the opposite side of the building, for example, will be obtained when this is the practice. Again, press photographs often are subject to this limitation.

After the Fire

When the fire has been extinguished and the firemen have ceased their strenuous activities, there is little need in the way of general photographs of the entire fire area. A few good record shots are desirable for general orientation, but little else is achieved by such photographs.

During the investigation, many points in the burn pattern, partially destroyed appliances, utility materials, and other items of possible importance will be noted. These should be photographed carefully enough to record their general features and *their relationships with other objects nearby*. This can be most important in determining whether there is evidence that the fire spread from one of these objects to other combustible objects or to adjacent structures. Such a fire spread must cast suspicion on the object that initiated it, even though it may have resulted from the fire rather than causing it.

The largest single activity related to most investigations after a fire is the clearing of the area so that critical points and items can be uncovered. In many instances this must be done by a crew of men, or even machinery, and may require days to accomplish. If possible, this is the time the investigator should be present with his camera, because he can obtain an immediate record of anything suspicious or informative that may be uncovered, before further clearing activities may destroy or damage the item in question.

It is expected that ultimately the actual point of origin will be revealed and, hopefully, recognized. At this point the investigator should carefully supervise the clearing and record photographically all of the evidence that gave rise to his choice of this as the point of origin. Whether he is a later witness in an arson trial or in a civil action, his photographs will be the best evidence of what he did, what he saw, and why he reached the conclusions he states as a witness.

Neither the human memory nor the notebook, important as they are, will ever match the camera as a means of recording observations.

Color vs. Black and White

Interior photography of structural fires is generally satisfactory in black and white, subject to the limitations discussed in the following section. But even interior fires may at times be better photographed in color because it gives a better correspondence to what the eye perceives. Most fire scenes retain little but blackened and unblackened areas, so this observation is only applicable to selected situations. Because of the added difficulties of color photography and the common need for color prints or projection facilities, it is generally sufficient to use black and white.

With outdoor fires in forests and grass lands, the reverse situation applies in many instances. Green and brown (from heat) leaves appear about the same with black and white film. The same is true of bark of trees and bushes. Here, color film is most desirable, even though black and white may also be used on the same investigation. It is desirable for the photographer to carry two cameras, one loaded with black and white for definitive photography and one loaded with color for general delineation of areas. While this is an added task to the many other things that the investigator has to do, it often pays dividends in the long run and is a well-advised action.

Late Investigation

In civil suits especially, it is not uncommon for an investigator to be assigned long after the fire has occurred and the site more or less cleared of fire debris. Photographs taken under these circumstances will necessarily be less useful than those taken under more favorable situations. To assume that they are not needed or will be valueless is an error. The most significant indications of the course and origin of a fire are rarely found in the rubble that is removed. Rather they persist as charred surfaces, deep burns, heavy discoloration, and similar features on the portions of the structure that are not hauled away. Late photographs of these features are often as valuable as if they had been taken at the ideal time. They are always desirable and frequently ciritcal in importance.

General Considerations

Fire scenes are some of the more difficult photographic subjects because of the properties of burned wood in relation to light. In addition, large structures offer special problems because most such scenes have insufficient light to focus a camera accurately, and flash lamps are ineffective at the distances that must be photographed in many instances. Another source of difficulty is that the general reflectivity of a burned scene is so small that it is recommended that the camera diaphragm be opened one stop, even compared with a meter reading. Otherwise, photographs tend to be underexposed.

Burned wood, although black, is highly reflective at certain angles of light incidence. These "glare" spots photograph white, appearing lighter in many instances than unburned wood nearby. The use of a Polaroid filter will assist in reducing these glare spots and giving a truer representation of color intensity. Also, critical areas may be reasonably reproduced by illuminating them head-on so as to avoid glare angles.

Illumination of regions too distant to be lighted by a flash is most conveniently accomplished by use of one or more "slave" flashes. These may be either electronic or bulb flash units attached to a photoelectric cell which actuates the flash when illuminated sharply by the flash unit of the camera. Such units are placed in the background and directed toward structures of interest. When the camera flash occurs, all slaves also flash from the tripping action of the initial flash. *It is important that the diaphragm stops be the same with as it is without supplementary slave illumination.*

When negatives are severely underexposed and cannot readily be retaken, a useful procedure is to enlarge or print them with a reinforced developer. Thus, ordinary Dektol stock may be used without its normal dilution and will force the prints strongly. Many times, this procedure will retrieve the situation when the negatives are so thin as to be unprintable by normal procedures.

INTERROGATION OF WITNESSES

The obtaining of statements from witnesses of a fire is an important function of the general investigator as opposed to the technical fire investigator. Both have an interest in statements, but the technical investigator should readily ascertain more about the cause and sequence of the fire than can possibly be obtained from witnesses alone. His interest in statements will generally be limited to essential background information. This is exactly the information that is often overlooked by the insurance investigator who is primarily concerned with the fire loss and the company liability; he often fails to ask the questions that might assist in elucidating the cause of the fire by technical means.

The matters indicated here as being important to ascertain from witnesses are limited to the type of information needed by the technical investigator. It is assumed that the financial and personal issues that are also of interest to the insurance investigator will be dealt with adequately by him.

Important questions that should be answered by all witnesses in order to assist the technical investigator with background information are as follows:

1. Where was fire or smoke first noted?
2. Where was the witness at the time of noting the fire or smoke?
3. What was the color of the smoke? (This *very* important question is rarely asked and often answered with conflicting statements.)
4. At the time of first noting the fire, were flames visible or was there only smoke?
5. From what part of the building was the smoke (or flames) emerging?
6. How rapid was the buildup of the fire? (Estimate minutes correlated with estimate of general involvment.)
7. Did character and color of smoke change during the fire and, if so, when?
8. Did any sharp noise or sudden eruption suggestive of an explosion occur at any time?
9. During what part of the fire did the explosion occur, relative to initiation and buildup period?

Witnesses should be questioned in the simplest and most direct terms possible. Left to themselves, the impression of the magnitude and destructiveness of the fire may alter their judgment and lead to exaggeration and even to neglect of the critical points, replaced by emotional reactions. Even with all precautions as to asking the right questions and careful note of all answers, it is general experience that stories of different witnesses will show frequent conflicts, sometimes seeming that they are describing different fires. This follows from the well-known fact that few persons are good observers. Added to this are the further facts that different persons will have different vantage points in space and different time intervals, and that they will be much influenced by remarks they overhear or that are made to them by persons trying to convince them of something they did not see or believe at the time.

In one instance it was critical to determine whether hydrocarbon liquid was the first fuel that burned in a large fire because arson was suspected. This would burn with a very smoky flame and generate large quantities of soot, a considerable amount of which appeared in several portions of the fire scene. It was thought useful to question the first arrivals at the scene, at the time when the fire was still relatively small. One declared that the smoke appeared white to him, a second said "very black," and a third decided it was "sort of gray." This fire occurred at night, which makes such observations difficult, but none of the witnesses hesitated to express a conviction as to the color of the smoke. It is clear that eye-witness statements are of variable reliability, with no means of distinguishing the reliable from the purely fictitious. Thus, the most important information to be had from witnesses concerns matters that are strictly nontechnical, and even here, it is necessary to use great care and skill both in interrogating them and especially in interpreting what they say.

Firemen on the scene tend to be restricted to the small area that is their responsibility and to be greatly impressed with the amount of heat, smoke, etc., all of which limits their ability to observe the more significant facts of the fire. This situation is common to all large fires and many small ones. The fire chief, who has overall direction, is usually the only one who may be expected to have much useful information. This follows from the fact that he is not limited to a small area of observation, but tends to study the fire *in toto* and from a variety of vantage points. His responsiblity is also greater, and so is his sense of that responsibility, so that he is likely to be generally more reliable than firemen under his command. At the same time, many fire chiefs, especially in small communities, are poorly trained and incapable at times of interpreting accurately what they see. Thus, even this source of witness information is not uniformly reliable.

It can be stated, as a general rule, that the technical investigator can learn far more about the origin of a fire by investigating the scene than by questioning witnesses. When he has general background information that is reasonably reliable, this should correlate with the information he can obtain from his own examinations. If it does not, the impasse may occasionally be resolved by further questioning. But usually the investigator is better off relying on the physical facts that he uncovers and, if necessary, upholding them against contrary statements of witnesses.

BODIES OF VICTIMS

It is not common for the general fire investigator to concern himself with the details of victims of a fire. In many instances there is nothing to be gained from such a concern, but in some instances it is well for the investigator to read the coroner's report on fire victims. There is a very definite reason for this interest.

There are two ways in which a fire may claim a human victim: (1) asphyxiation from carbon monoxide and (2) burns of the body itself. Of the two, No. 1 is much more common than No. 2. A person sleeping in a second-floor bedroom with a fire burning on the lower floor has a poor chance of avoiding asphyxiation which is liable to occur long before flames invade the bedroom. This type of situation is not at all uncommon. In order for a person to die directly from burns, he must be physically *in the fire* while still alive which is far less common than his being inside a structure which is burning.

A proper investigation by a coroner will always include analysis of the blood for the percent of carboxyhemoglobin contained in it. If that quantity is greater than 40 percent, it is highly probable that the person was at least unconscious at the time the fire reached him. If it is 60 percent or more, it is certain that he actually died from the fumes, regardless of the presence or extent of burns.

This information is of utility to the investigator because it can be considered in connection with the remaining information to determine something of the general sequence of events and the activities of persons in the burned structure. An arsonist, for example, might receive severe burns from some accident or miscalculation, but he is most unlikely to die by asphyxiation. Those persons who died of burns were in the fire; those who died from carbon monoxide were away from it but subject to permeation of the environment by it. Persons on a floor above an active fire are likely to die from asphyxiation, because the hot lethal gases travel upward. Persons on the same floor with a fire have a far less hazard from this source. Considerations such as these will often be of value in the investigation and should not be overlooked.

LABORATORY EXAMINATIONS

Detection and identification of volatile flammables are the most important tasks of the laboratory examiner but not the only ones, nor even necessarily the most frequently needed ones. This type of examination will not generally be carried out by the general fire investigator, although he may also be skilled in laboratory methods. Detailed procedures will not be given here—rather, a summary of the various operations that may be performed.

Isolation and Identification of Accelerants

The normal manner of isolating a liquid accelerant from other materials, generally solids such as soil or wood fragments, is to distill the liquid from the solid residue in a current of steam. Equipment to be used for such a distillation must be large enough to accommodate considerable amounts of miscellaneous solids, and the vessel must have an opening for filling that will allow relatively large items of material to be inserted. Various designs are used, sometimes specially made for the purpose. A water boiler and steam inlet to create the steam current through the material and a condensing system complete the outfit. Vacuum distillation in a large dessicator has also been described by Nichol (1). Supercooled traps serve to condense the vapors.

Separation of volatile liquids under these circumstances necessitates distillation of considerable water. When the accelerant is not soluble in water, it will appear as a separate layer in the receiver, generally floating on top of the water because virtually all useful accelerants are lighter than water. Some such liquids are miscible or soluble in water. These would include any lower alcohol, acetone, and a few other materials. No separate layer would be noted in case this type of liquid had been employed, and separate tests would be required to identify its presence in the water distillate. Sometimes it will be distinguishable by smell as a characteristic odor. Partial separation can be obtained by subjecting the distillate to further distillation, preferably with a fractionating column. Such a fractional distillation apparatus is also useful in studying hydrocarbon mixtures to get an idea of their components, boiling range, and similar factors.

Identification of the accelerant may not be simple because mixed materials have generally lost much of their volatile content before being recovered. By running distillations, occasionally chemical tests, flash points, and refractive index or density, it is possible to distinghish between most of the possible mixed materials. A better procedure is to employ the gas-liquid chromatograph, which divides the liquid into its separate components and fractions. Comparison of the resulting composition curve with those of possible materials, e.g., gasoline and kerosene, will serve for a reasonably positive identification. In fact, because there are differences between batches and lots of such materials as gasoline, it is sometimes even possible to determine with some reliability whether or not the material came from a particular suspect source (2).

Numerous physical methods of analysis have also been employed, including infrared absorption spectrometry and mass spectrometry, both of which have a great capability in the field of identification. Other physical methods include determination of refractive index, density, and boiling point or boiling range. These methods are generally within the capability of the crime laboratory.

From all of these laboratory procedures, the most important single piece of information that is made available is that a foreign flammable liquid was present at the fire scene. This alone is strong evidence for arson, at least after the possibility of accidental placing of the liquid is eliminated. When it is possible to obtain still more specific information, it is evident that further conclusions may often be drawn, and a more complete investigation has resulted. It has been possible at times to match the properties of recovered liquid fuels so closely to those of residual liquid in a container as to create strong evidence that the container was the source of the fuel. This is more likely to be possible in an aborted arson, in which ignition has failed so that the fuel remains unburned.

Other procedures for separating liquid accelerants from solid debris have been attempted with limited success. Extraction with a very volatile solvent could be used, but the method is neither easy nor economical. One possible shortcut that probably could be used, although to this writer's knowledge has not been, is to enclose the debris in a closed container, heat it somewhat to volatilize the liquid and saturate the internal gaseous phase, sample the vapor, and analyze it by gas chromatography. This procedure would require special equipment also and would not be expected to detect some of the less volatile fractions of mixed fuels. It would, however, in all probability be successful in many, possibly most, instances.

A procedure very similar to the one suggested has been recently published by Ettling and Adams (3) who carried out experimental recoveries of accelerants in a controlled system and used the gas chromatograph for the analysis of the vapor phase following burning. The method, applied to a variety of possible accelerants which were placed on several types of materials and burned, appeared to yield very useful results.

Examination of Burned Materials

It is often advantageous to submit various types of burned material for laboratory examination. Determination of the type, extent, and character of the burn, and frequently of the type of material burned, can be made in the laboratory; it will rarely be as totally satisfactory as a field examination. The materials that might require such examination are manifold, including items of wood, metal, cloth, paper, paint, roofing, and other products.

Some examples of such examinations that are possible by use of low power microscopes, measuring equipment, and special photographic methods follow.

Burned wood fragments, taken from selected areas, will allow measurement of depth of burn at these points; character of the burn which may be significant in terms of the conditions existing; and possibly of the type of wood burned. Unburned portions of the fragment may be studied as to the species of the wood if this contributes anything to the interpretation. Burned wood fragments from unknown positions prior to the fire may also be studied to ascertain where they came from and the nature of the wooden object burned. This can be accomplished either by study of the form of the fragment or of the type of wood. Larger pieces may be studied as to the burn pattern existing on them, although this information may generally be obtained in the field investigation as well as in the laboratory.

Fused and oxidized metal may well require laboratory examination. If the fire is possibly to be ascribed to shortcircuiting of electrical wiring, it will be highly important to photograph fused areas under conditions that are more suitable than those existing in the field. It may be desirable to examine such materials for presence of metals other than the (copper) wire normally used, e.g., to determine if zinc or other low melting metal is alloyed with the wires. The degree of oxidation can better be determined in the laboratory than in the field; the condition of residual insulation, its type, and the extent of deterioration are best examined in the laboratory.

Suspected failures of appliances often require examination of critical parts under laboratory conditions. For example, a critical electrical component or a defective area in a burner cannot generally be thoroughly tested in the field but can be removed for a detailed examination. The possibilities in this area are so extensive and variable that only an indication of their importance can be included here.

Burned paper and cloth will rarely be of consequence in general fire investigation, but are common items in bonfires and fireplace fires used to destroy such materials. At times, it is highly desirable to determine the nature of the material so destroyed. The distinction between burned paper and burned cloth is obvious to the eye as a rule, but it is not so well known that often the type of paper, or of cloth, can also be determined. This examination starts with microscopic (low power) study of the material, taking into account the fact that texture (especially important with cloth) is generally retained in the charred material. Further tests can be made by continuing the ashing of a little of the material until a white or uniformly gray ash is obtained and then comparing it with known standards of burned similar material in gradient tubes (4).

Another area of concern with burned paper especially is the decipherment of printed or written material on the charred sheet. This also is possible by laboratory methods in nearly all instances. Here, however, it is highly critical that *the field investigator*, who is not familiar with the proper method of collection, *leave this task to the laboratory worker* who will continue the decipherment. Any inclination to rush into a problem such as this generally results in the charred material being broken up so badly as to prohibit further study. The most critical phase of the entire process is the collection of the material, which must be done in such a way that the charred material does not become fragmented.

Special and Miscellaneous Examinations

Every crime laboratory is equipped to carry out a number of special types of examinations and experimentations that are not otherwise available to the fire investigator (5). Some of these involve burned material as discussed above, and others are concerned with extraneous, but often very important, types of laboratory procedure. These are discussed below.

Breaking and entering. It is not uncommon for arsonists to enter premises illegally and by means common to those used by burglars. The point of entry may not be burned and, in fact, has a good chance of not being burned because its easy access to the arsonist also makes it accessible to the firemen who often protect it unwittingly. Burglars (and arsonists) may leave tool marks where a door or lock is forced, and these can be connected with the tool that made them. Windows may be broken, in which case the person breaking them will acquire glass fragments in his clothing. Sometimes, it may be possible to connect such fragments with the broken window, provided the effects are not obscured by the effect of the fire on the window. Screens may be cut, and metal fragments, as well as fragments of paint, wood, and similar trace materials may be collected by pant's cuffs and pockets of the person who broke into the premises. The possibilities here are numerous and rarely utilized fully in the investigation of fires.

Fingerprints. It is not common to locate fingerprints at the fire scene or on materials used by the arsonist. It must be suspected that this fact may be due more to the lack of looking for them than otherwise. Clearly, if the fingerprint is exposed to the actual fire, it will be destroyed, but it is often possible to locate containers that have not been exposed to direct heat by the fire to an extent that should be destructive. Likewise, exteriors of points of entry may carry prints and be unburned after the fire. While it is not expected that fingerprints will often contribute greatly to the solution of an arson, it is enough of a possibility to warrant awareness of its utilization when indicated as a possibility. Attention to some of these opportunities for definite identification of arsonists has been called also by Steinmetz (6).

Special experimentation. It is inevitable that the field investigator will encounter many situations in which he has too little information to be certain of an interpretation that he is inclined to make. *In such a situation, controlled experimentation is essential.* This can generally be done only under laboratory conditions, and if due consideration is given to the differences between large and small fires, such experimentation may be very well accomplished with small fires rather than their larger counterparts. The properties of the materials can be chosen to be identical and the principles must be the same; therefore, the results will be as applicable from the small experimental fire as from the holocaust. Some illustrations of the type of experiment that may be performed will be included in Appendix 1. Such experimentation is one of the prime functions of the laboratory in the investigation of fires.

References

(1) Nicol, J. *News Letter, International Assoc. of Arson Investigators* Feb. 1950.
(2) Cadman, W. J. and Johns, T. "Application of the Gas Chromatograph in the Laboratory of Criminalistics." *J. Forensic Sci.*, 5, (3), 369, 1960.
(3) Ettling, B. V. and Adams, M. F. "The Study of Accelerant Residues in Fire Remains." *J. Forensic Sci.*, 13, (1), 76, 1968.
(4) Brown, C. and Kirk, P. L. "An Improved Density Gradient Technique and Its Application to Paper and Cloth Ash." *J. Crim. Law, Crminol. and Police Sci.*, 43, 540, 1952.
(5) Kirk, P. L. *Crime Investigation, Physical Evidence and the Police Laboratory*. Interscience Pub., New York, 1953.
(6) Steinmetz, R. C. 1952 Annual Conf., International Assoc. Identification, Havana, Sept. 8-11, 1952.

Supplemental References

Bennet, G. N. "The Arson Investigator and Technical Aids." *J. Crim. Law, Criminol. and Police Sci.*, 49, 172, 1958.
Brackett, J. W., Jr. "Separation of Flammable Material of Petroleum Origin from Evidence in Fires and Suspected Arson." *J. Crim. Law, Criminol. and Police Sci.*, 46, 554, 1955.
Burd, D. Q. "Detection of Traces of Combustible Fluids in Arson Cases." *J. Crim. Law, Criminol. and Police Sci.*, 51, 263, 1960.
Chanmugam, W. R. "Scientific Investigation of Origin of Fires." *Ceylon Assoc. Sci.*, Sept. 1946 Sessions, Presidential Address, Colombo, 1947.
Kennedy, J. "Photography in Arson Investigation." *J. Crim. Law, Criminol. and Police Sci.*, 46, 726, 1956.
Schuldiner, J. A. "Identification of Petroleum Products by Chromatographic Fluorescence Methods." *Anal. Chem.*, 23, 1676, 1951.

13

Arson

Arson is defined as: "The malicious burning of a dwelling house or outhouse of another man; the similar burning of other property, including one's own house."* To be a criminal act, *intent* is a necessary component. Otherwise, the fire is classified as accidental. Virtually all persons kindle large numbers of fires throughout their lifetimes—smoking materials; open fires in fireplaces, the bonfire, or the barbecue pit; even the lighting of a gas stove or furnace constitutes the setting of a fire. Even when intentional fires in some manner get out of control and develop into a large fire that is uncontrolled but still unintentional, no act of arson is involved.

Aside from the matter of intent, there are other differences in the fires themselves that serve as a basis for the determination of the fire as resulting from an act of arson. However, the method of investigation of the arson fire is not basically different in any respect from that of the fire that results accidentally. Furthermore,

**Webster's New Collegiate Dictionary*, G. and C. Merriam Co., Springfield, Mass., 1961.

lacking any real basis for decision between the two, *every fire should be investigated as though it could be the result of arson.*

Even though the major investigation of a fire should be in the physical remains, there are other considerations that may be important to the investigator, involving the mental processes or psychology of the arsonist, his motives, and his responsibility. These will only be touched upon briefly in this volume, but they are too important to be totally dismissed.

THE ARSONIST

Although nearly anyone might become an arsonist under certain circumstances, those who actually do take up this activity can probably be classified under three headings:

Arsonists for profit. This includes perhaps the largest single group of persons who burn their own property for the sake of collecting insurance, or who will burn another person's property for hire. In this context, arson is a calculated act not basically different from burglary or armed robbery, although its perpetrators tend to feel less criminal and more justified by the unfortunate aspects of their individual financial situation, and the fact that only an insurance company which is a large and rich corporation will stand the loss. Such persons will certainly meet all the legal qualifications of felons, although to place them formally in this category is often very difficult.

Arsonists for spite. Persons in this category are "getting even" or seeking revenge. Someone has wronged them, the wrong being either real or imagined, and the most obvious recourse to these warped minds is to burn the property of their persecutors. Arson, in this context, is undoubtedly more a rural than an urban crime, but no less a crime, and a serious one.

Arsonists for "kicks." This includes two different categories of personality, the first being the "firebug" who has a pathological attraction to fires and is happiest when witnessing a fire and its destructive effects. This person is distinctly abnormal and should be subject to psychiatric rather than penal care, because the glaring quirk in his personality will not be cured by simply imprisoning him. The firebug is a very dangerous person, but he is also more subject to recognition than most arsonists, because he will invariably be at the scene of the fire, obviously relishing it. In addition, he will tend to set several fires in succession, attending each event in a conspicuous location. His apprehension is generally a police problem, but his responsibility is often not simple to prove, largely because of the independent and frequently uncoordinated approach of police and fire fighting personnel.

The *second* category of arsonists for "kicks" is the rather casual but malicious prankster who sets fires merely for the momentary excitement or as a general retaliation against society. This person, generally youthful and delinquent in more ways than one, has no special attraction to fires as such, but he will also do some thievery, release car brakes on hilly streets, break street lights, and possibly beat old persons, just for "kicks." He is also a very dangerous person, not specifically as an arsonist but rather as an uncontrolled and irresponsible trouble maker. His apprehension is therefore much more difficult as relates to the arson itself, because his behavior is erratic, unpredictable, and follows no distinct pattern. He is also a pathological case, but one with less definite distinguishing characteristics that could simplify in some measure his treatment by psychiatric or other means.

It is evident that the investigator of physical evidence of the fire is concerned very little with the type of arsonist who may have set it. However, there are differences in *modus operandi* which he may note in the investigation, and these can be of great help both in tracing the arsonist and in producing information useful for trial purposes.

A major difference in the types of arson that will be apparent is in the matter of access. The owner of a building will have, ordinarily, complete access and familiarity with the premises; he will therefore show a great deal more care and thoroughness than is likely to be exhibited by the other arsonists listed, except perhaps for the paid professional. When multiple origins are present, and especially if doors have been opened without signs of breaking and entering, suspicion must fall on the person(s) who had access and familiarity. The *third* category represents the opposite extreme, where the fire may be started on the exterior, a Molotov cocktail is thrown in through a window, or access has obviously been gained by breaking a door or window. The latter situation is perhaps even more consistent with the hired arsonist, who wishes to be very thorough and who is likely to understand the requirements for effective arson. Revenge arsonists also may fall into this intermediate group, but they are more likely to be classified with the more casual hit-and-run types who set fires for fun. An unusually malicious motive, for example, cannot be well deduced from the physical evidence in most instances, but when there is unusual emotional content to the act of arson, there are often other acts of wanton destruction which accompany the fire. All of this type of information, when obtained, may be considered as a bonus by the investigator of the physical evidence.

THE ACT OF ARSON

The determination that a fire has been intentionally set is central to all fire investigation and is one of the more difficult phases of that investigation. To understand the procedure for such an investigation, it is first necessary that a well-balanced understanding of the possibilities and difficulties of building a fire be thoroughly understood. Nearly all intentionally set fires are built for a specific purpose. Yet the problem of building a fire to warm oneself is not basically different from that of burning a building for purposes of excitement, revenge, or the collection of insurance. The chief difference resides in the ultimate purpose, which calls for *fires of different magnitude*. The fireplace fire will be small and controlled, while that set by the arsonist will ultimately be large and destructive. To achieve the latter aim requires that the fire build up rapidly at some stage of its development. In contrast, it may be considered desirable that its initial stage be slow and inconspicuous so that the person who set it may escape and avoid suspicion. Thus, the arsonist must adhere to good principles for the starting of fires, and he has a most difficult problem in controlling the speed of development of the fire.

It has been noted for a long time that the investigator of crime is most effective when he can place himself in the role of the criminal; the best investigators are those who can do this most effectively. They can learn to think as the criminal thinks, react as he reacts, and from this can estimate how he operates. This generality is no less true of arson investigators than of those who investigate other crimes. An effective arsonist has had experience with fire and a feeling for its behavior. He also must calculate his results well. When the fire investigator is as aware of the principles and generalities of fire kindling as the arsonist whom he is trying to apprehend, and he can place himself in a frame of mind similar to that of the arsonist, he will be an effective arson investigator. He must first understand the basic requirements for successful setting of fires.

As discussed earlier, the fundamental requirements that must be met are the *ready availability of fuel, proper ventilation*, and *ignition*. While much attention has often been given to the ignition devices of the arsonist, it is reasonable that the more important considerations for success of the fire reside in the matter of fuel and access of air. Only when delayed ignition is essential is there any need for unusual methods of igniting the fire. It is certainly true that by far more arsonists choose their time and circumstances to allow their escape from being revealed than rely on any unusual type of igniting device. Perhaps the preoccupation of some investigators with such devices results primarily from their infrequency which adds much interest to their discovery.

Arranging the Fire

The arsonist, having determined the structure which he intends to destroy, must make several choices that are critical to his success. He must choose a point at which flames will rise into fresh fuel. Although most fires spread maximally under ceilings, it is evidently impractical to start the fire directly under a ceiling. Thus, a floor adjacent to a flammable wall is one of the likely points for initiation of the fire. The frequent use of sheetrock (gypsum board) as wall coverings limits seriously the availability of walls that are suitably flammable. It also generally prevents serious penetration of ceilings. *Furniture* in suitable arrangement and of suitable type is another possibility, but this has several limitations, chiefly that the furniture may burn but not spread the fire rapidly to the structure. Also, some furniture is very unsuited to rapid and effective burnings. Thus, older davenports and similar items were often stuffed with cow hair held in place with burlap. Burlap burns well, but is usually present only in small amounts; hair, present in large amounts, burns very poorly. The more modern foam rubber stuffing is excellently suited for burning, because once ignited, it burns with an intense flame and effectively increases the fire volume.

Basements are a very suitable place to initiate an act of arson. Here, the walls are generally not finished with fire-resistant covering; the entire building structure is above the flames, thus providing the maximum quantity of fuel; and concealment of the fire during its early stages is maximal. Any basement fire should be investigated with special care, because of all possible locations, this one is generally most suited to the arsonist's purposes.

Attics are the least satisfactory for initiating large fires, and attic fires are therefore less suspicious. The location is unsatisfactory for the opposite reasons given for the suitability of basements. A roof may be lost, but the resulting fire is wholly unsatisfactory as a means of destroying an entire building in a holocaust. One possibility, that the arsonist has access to an attic and has had little experience with kindling fires, may lead to his choice of this location. Such a person would ordinarily not be in the class of firebug, but might be attempting to collect insurance. The frequent accumulation of much flammable trash in an attic could provide a temptation to choose this site.

Wardrobes are not infrequently the location in which a fire is set. The choice may rest on the rather large amount of available fuel in the wardrobe which is filled with clothing, boxes, and other combustible materials. Another point in its favor is that the control of ventilation is easy, and the fire may burn at a low intensity, undetected, while the arsonist makes his escape. Eventually, the fire will erupt from the wardrobe, already well developed, and may then engulf the entire structure. In dealing with fires set in wardrobes, it should be remembered that woolen clothing is not good fuel, that a fire may extinguish from lack of air in a closed wardrobe, and that leaving a door open for ventilation will lose the advantage of late discovery.

Fireplaces, Furnaces, and Heaters

Arsonists experienced in fires and firefighting are likely to kindle a fire in the vicinity of a heater, fireplace or similar area in order that the fire may be termed accidental and connected to the malfunction of the utility. For this reason, any such fire should be examined with special care because amateur fire fighters and investigators may well fall into the trap set by the arsonist.

The key to uncovering this situation is again a study of the pattern. In most such instances, the arsonist will use a liquid flammable which is poured around and under the facility and then ignited. In such instances, it is normal for the liquid to penetrate well below the utility in question and to give a burn far below that which is possible for the utility itself. Both gas appliances and fireplaces would normally lead only to fire above the base level of the normal combustion, not to a region under the floor that supported it. Burns below this level of support are almost always traceable to the use of an accelerant. In some instances there may not be penetration of the floor under the appliance. However, there will still be burning clear to the floor level, which again is not normal for a fire escaping from an appliance unless there is very obvious and severe damage to the facility itself. Naturally, the investigator will need to check carefully that no such abnormality in the appliance exists. An appliance without a severe abnormality (see sections on appliances, fireplaces, etc.) is not expected to start fires under any ordinary circumstances, and any fire in such a vicinity is to be considered suspect and investigated with extraordinary care.

A house occupied by a fireman suffered a fire in the living room in the neighborhood of a metal gas-log fireplace. Although severe damage was done to the room, the house was not consumed due to efforts of the firemen. In such an instance the gas fireplace was obviously suspect and might have been assumed to be subject to some malfunction. The floor was raised somewhat above the underlying earth, and the supporting joists under the fireplace were found to be heavily burned *from the bottom upward*. This could only have occurred from a fire at the level of the ground and certainly not from any malfunction of the gas equipment, all of which were found to be in operating condition and without defects. In such an instance, there is only one possible conclusion, as outlined above. Such occurrences are not infrequent, especially when the owners have a good appreciation of fire behavior, as is true of some firemen, and of persons who practice arson on a professional or semiprofessional basis.

Trash Accumulations

Trash accumulations that are apparent as a portion of the fire scene will ordinarily be scrutinized very carefully, because they provide a suitable place for kindling of fire, especially when saturated with a liquid accelerant. On the other hand, they also provide an inherent fire risk from accidental causes. Both possibilities must be carefully weighed before considering that a trash pile has been used by an arsonist. If liquid flammables have been poured on the trash, it is highly probable that some of the liquid may be recovered and identified, in which case, strong evidence of arson is available. When the trash pile appears to be an origin or possible origin of the fire, it is highly important to search for flammable liquids. It is also important to search for other means of ignition that could be accidental, because trash does not normally ignite itself but may be ignited by accident as well as by design.

Such trash accumulations are frequent in basements, in attics, in rooms used for general storage, and around back doors. A trash-filled basement is ideal for producing a large fire, and arson is not uncommon in such quarters under business establishments especially. Of all fires, these are perhaps some of the most difficult to investigate, because in such a basement there is nearly always some possible means of accidental ignition, such as defective electrical wiring, furnace, water heater, or similar device or condition. Exterior trash accumulations are often utilized by the firebug type of arsonist who does not have normal access to the interior of a structure. Since they are generally discovered promptly, such fires do not ordinarily have time to penetrate into the interior of the building, and they are likely to produce little damage.

All of the above fuel sources may be considered as preexistent on the premises burned, and they may sometimes be considered as being of less consequence than an accelerant such as a liquid flammable which is applied by the arsonist. Such an assumption will be only partially true, because however well the liquid flammable burns, if it does not involve the structure properly, a very poor result—from the arsonist's standpoint—is produced. Because of all the uncertainties attendant on kindling a fire with preexistent materials, few arsonists choose to rely on this source of fuel exclusively. Liquid flammables, or accelerants, are characteristically found in a very high proportion of arson fires.

Liquid Flammables

From the arsonist's point of view, the placement, quantity, and type of liquid flammable are very critical. Although such a material has the excellent property of producing much fire rapidly in any given region, there is no assurance that the flames so rapidly produced will find other fuel in their path so that they can build into a conflagration. The larger the quantity of flammable material used, the more rapid and general will be the resulting fire. However, it is difficult at best to transport and arrange more such fuel than can be easily carried by hand, except in certain favorable circumstances. Thus, the quantity is unlikely to be measured in units larger than gallons at most, and often in quarts or less.

A mammoth fire occurred which involved approximately an acre of storage ground for used and worn-out tires. The fire developed rapidly and was of extreme intensity. At the edge of the fire area was a truck carrying a number of fifty-gallon drums. The truck had been consumed to the furthest extent possible along with the tires. It is apparent that under some circumstances, and especially in isolated regions, the rule of using only the amount of accelerant that can be carried in the hands is not valid. Tires can burn fiercely when the fire is in progress, but they are exceptionally difficult to ignite. Without outside assistance, it is almost certain that the tires could have remained as they were stacked indefinitely without any fire hazard at all. A fast developing fire under these circumstances certainly requires an explanation, which seemed to be provided by the truck with the drums still loaded on it.

Placement of the liquid for maximum results is often difficult from the point of view of the arsonist. Most common is a pile of papers or other trash on an open floor or over furniture, or some combination of these. In a basement, such placement generally allows some fuel to soak into trash or into earth and escape combustion. A pile of papers soaked in gasoline, for example, allows fire to burn over the top of the pile, the papers feeding additional fuel to the fire in much the same manner as the wick of an old-fashioned oil lamp feeds kerosene to the flame. Portions of fuel in the interior of the paper pile are not heated above the boiling point of the fuel at most, and often so much less that the fuel may be found even days or weeks later. Fuel soaked into the soil likewise is poorly consumed, and only on the surface, so that residues of it may be found and identified after considerable periods. Parker, Rajeswaran, and Kirk (1) poured gasoline on the ground and tested the retention by the soil. Within hours, the more volatile fractions had been lost, but the less volatile persisted for a considerable time, depending on conditions. The remaining vapors were readily analyzed by gas-liquid chromatography.

Liquid fuel placed on a floor is very common and frequently leads rapidly and certainly to the proof of arson. This is partially true because of the lack of correct information about the properties of the fuels, as discussed earlier. On a tight floor, little or no damage to the floor may result. On a floor with cracks, holes, or joints through which the fuel may penetrate, holes and lines will burn into the floor in a manner very characteristic of liquid fuels below the floor surface. Some floors are just a short distance above ground level; if fuel penetrates, it may soak into the underlying soil, with all the consequences detailed above. Another common mistake the arsonist makes is to place the fuel in the center of a room without ensuring that the flame will reach furniture, walls, or other structural units which are necessary for the proper spreading and development of a building fire.

Multiple site placement is common in arson. Not satisfied to gamble on a single point of origin, many arsonists place the liquid in various locations, so that the resulting fire will build much more rapidly and widely than from a single region. It is this tendency that gives rise to the axiom of fire investigators that when there are multiple points of origin, it is virtual proof of arson. As a practical matter from the standpoint of the arsonist, the choice of multiple sites can be a double-edged sword. It obviously requires more time to distribute fuel in several places than in one, and while later distribution is proceeding, vapors from the earlier points may be forming an explosive environment which may well involve the person of the arsonist. If he ignites each when it is deposited so as to avoid explosion, he then is carrying a dangerous quantity of fuel in his hand, which may well involve him in a fire. His clothing often receives spatters of the fuel and may ignite. Especially if the fuel has a high volatility, the hazards from these sources are likely to outweigh any advantage that is gained from multiple distribution of the material. In addition, the earlier small fire may be spotted and his risk of being caught is greatly increased. A low volatility fuel, such as kerosene, is suitable for this type of operation, but one like gasoline, or even paint thinner, can be extremely dangerous.

The "molotov cocktail," a bottle of gasoline, in its common form has long been used by some arsonists and is currently quite popular since it can be thrown and thus allows for a rapid escape. If it is not aflame when thrown, it must depend on some source of ignition present in the area exposed to gasoline from the broken bottle. It can be rigged in such a manner so that it is afire when thrown, thus carrying its own ignition source.

Arson from this source is not difficult to detect because of the accompanying phenomena, a broken bottle, possibly a broken window, and often eyewitnesses—possibly from a passing car. Tracing it to the arsonist may be very difficult but often is not. It must be remembered that the broken bottle will be at the bottom of the fire, where it is exposed to minimum heat, so fire damage to the fragments is minimal. Thus, the investigator may be successful in collecting much or nearly all of the bottle fragments. These will be investigated by conventional criminalistic methods, including search for fingerprints and their development if found, as well as reconstruction and identification of the bottle, which may in some instances be traced due to special characteristics revealed by the investigation. In cases where no such fortunate evidence is available, other means, unrelated to conventional fire investigation, will have to be used, sometimes without success.

Plastic fuel containers filled with fuel have been encountered in arson cases. These can be placed over room heaters so as to soften and melt from the heat, thus spilling the fuel contained in them and producing a large fire rapidly. This device will delay ignition, but it also leaves residues that provide clear proof of arson and a possible lead in tracing the arsonist.

Quantity of fuel. As indicated above, the most likely quantity of liquid fuel is the amount that a man can conveniently carry, with the further stipulation that he may wish to attract no attention while carrying it. It has to be accepted that a man can easily carry an amount sufficient for kindling a very large fire, so this is not necessarily a serious limitation, except where fire-fighting facilities are likely to be available very rapidly. The burning of a large business building within a block of the fire station for example, would require much extra consideration and planning as compared with a residence on the outskirts of town or in the country.

For such a difficult operation, it would be desirable to have a rather large reservoir of liquid fuel available over a period of time. It would also be desirable that this fuel be immiscible with water so that it would not be extinguished by the hoses of the firemen. Any hydrocarbon fuel will meet this specification. One manner in which such a reservoir can be provided is by placing a metal drum of a suitable fuel such as gasoline, paint thinner, or other rather volatile material at the site of the origin, and placing enough fuel under or around it to start the fire and to heat the drum to the point of rupture. A well-filled drum is not difficult to burst by heating, and the ruptured drum will then spray fuel into the fire for a long time, with virtually no possibility of extinguishing it by conventional water or fog sprays. A possible undesirable side effect is the possibility that the rupture will lead to a large explosion by suddenly mixing imprisoned vapors under pressure into a fire. However, this is not usually undesirable from the arsonist's viewpoint, provided that he himself is not too close to the explosion. Such a drum placed at the origin of a fire kindled with additional fuel is not likely to be suspect as an instrument of arson, because there are many legitimate reasons for storage of liquid flammables on most premises.

Type of fuel. Of all the problems the arsonist has, this is one of the more difficult. Procurement of quantities of liquid flammables other than gasoline for his automobile or tractor may be remembered by the seller. If he intends to paint, a quantity of paint thinner is normally bought, and this could provide a cover. Kerosene, from many standpoints, is one of the best accelerants, but few persons have logical reasons for buying it in quantity. Industrial alcohol, either wood or grain, may be justified for several purposes and not raise suspicions. For example, some duplicator fluids are composed mostly of methyl (wood) alcohol, and a supply for the office is normal. From previous discussion, it will be noted that alcohols leave something to be desired as compared with hydrocarbons. Another common and available material is turpentine, but again, this is normally bought in small quantities only. Various other materials are normal for some individuals, but the average arsonist has only limited possibilities that do not arouse any suspicion. Gasoline will generally be the most readily available material, and this liquid does appear commonly in arson.

The act of setting a fire with a liquid flammable requires consideration of still another very important factor—the volatility of the fuel. *Gasoline* is relatively volatile and will create explosive conditions in the environment rather rapidly. If it is to be used, it is not well for the arsonist to linger too long in the neighborhood where the material has been spread, nor is it safe for him to kindle it from too close a range. These factors limit the extent and ease of the distribution and can well give rise to unexpected difficulties.

Kerosene on the other hand, is quite easy to use because it has low volatility and a relatively high flash point and it is unlikely to cause any explosion. On the other hand, any unburned traces of it that survive the fire will continue to survive for a long time, making definite proof of its presence possible.

Alcohols, even more volatile than gasoline, carry the same types of hazard to the arsonist. When somewhat diluted with water, they will still burn, but with much lower hazard as well as effectiveness. Liquors of ninety to one-hundred proof are capable of being used as fire accelerants, albeit not the most effective ones. Most of the other possible liquids will be relatively restricted as to availability, with the major exception of *paint thinners*. Hydrocarbon thinners are of the naphtha group and contain less highly volatile fractions than gasoline, but they are somewhat more volatile than kerosene. They are commonly encountered in arson investigations. Other *naphthas*, used widely in industry and in dry-cleaning, are suitable fire accelerants. Generally, their properties lie between those of gasoline and kerosene, as already discussed.

The above considerations are of consequence to the arsonist in determining his procedure in setting the fire. They are of equal consequence to the fire investigator who must correlate the observed facts about the fire with the possibility that it was intentionally set. They also serve as a guide to the thinking of the arsonist which is related to his experience in setting fires. An inept arsonist may well set himself on fire, cause unnecessary or premature explosions, or kindle a fire which is not truly self-sustaining or capable of development into a large conflagration. These things, and even the means of procuring liquid fuel which may often be traced, all serve as indicators of the possible identity of the arsonist and his previous actions.

Method of Kindling

The method chosen by the arsonist to kindle a fire will be related closely to the type of fuel and to his own thinking as to avoidance of suspicion. If he chooses to utilize liquid accelerants, he is limited to nearly immediate ignition because some of the vaporizing fuel may be lost during the time of delay, as well as create explosion hazards. With less volatile materials, the time of delay may still be considerable, but it is impossible to leave delayed flaming devices such as a lighted candle. Inevitably these will not delay the fire for long.

If the primary material to be ignited is not volatile, the delayed action ignitor is suitable, but in general the development of the fire will be less certain and satisfactory. Here again the dilemma that must be faced by any arsonist is to avoid the effort to combine ignition methods and fuels when the two are not totally compatible. While very ingenious procedures have been worked out to achieve the aim of the arsonist who has planned carefully, it is unquestionably true that most arson fires are set in a hurry, with means that are readily available, and without either elaborate planning or execution.

The choice of primary ignition devices may be broken into categories as follows:

Flames. Simplest of all to produce, the flame of a match, a lighter, or a faggot is easy to obtain, to apply, and to eliminate as evidence. The flame is a certain means of starting the fire, since it is a fire already and merely requires propagation. Whatever is left at the scene, e.g., a burned match, is likely to be destroyed to a point of escaping detection, although exceptions have been noted. One important exception is the candle which normally is not well enough destroyed to remain undetected. Melted wax will persist in a fire, even though the form of the candle which was its source is lost. As a delaying device the candle has received some attention, but probably more as a fictional than a real method. In the unlikely event that such a device is found after the fire, and no explosion has occurred to initiate the fire, it is quite certain that liquids of a volatile nature have not been employed. Even liquids of very low volatility probably have not been used; in fact, such liquids are relatively ineffective as accelerants although they may add to the intensity of the fire. Fuel oil, for example, is extremely difficult to burn except under conditions of high dispersion and air admixture. When applied to paper or trash, it can burn vigorously and add to the fuel, but it will not be readily ignited.

In one experiment, fuel oil was poured on a concrete slab and efforts were made to ignite it. Not even a small bonfire built on it was successful. A blowtorch applied directly to the fuel was effective in burning the oil around the outer ring of applied flame, but this fire extinguished spontaneously when the flame was removed. The major effect of the flame was to drive the oil away from it at the point of contact of flame with concrete. This test was made necessary by a claim that fuel oil on a concrete basement floor had been ignited by a flash-back from an oil furnace. It is clear that even a flame must be directly applied to an easily flammable material in order for a fire to result, although the variety of flammable materials that can be ignited by a match or lighter is greater than is true of some other sources of ignition.

A special case is that of the flame thrower, weed burner, or similar device that generates and maintains a large and hot flame for a protracted time. The use of such a device is not common but is highly effective, even with relatively refractory fuel. It is possible to ignite even quite large timbers and to spread a fire widely through a variety of flammables with so large and hot a flame. In one sawmill fire, it was determined that a weed burner was the ignitor, because of the extensive deep burns on the bases of heavy timbers that were lying on soil where no other source of such heavy flame was possible. The burner was also located close to the scene. Not only was fire started in this instance by the burner, but it was started in a number of places, so that the overall development of the fire was very fast and the destruction extensive.

Smouldering materials. Although mattresses, sawdust piles, and other fuel materials with limited access to air may be afire, smoulder, and ultimately build into a large fire, these are not favorite ignitors for arsonists nor are they primary, since they also must be ignited by something else. Their advantage for arson is in delaying the outbreak of the fire, allowing time for escape. This advantage is more than offset by the unreliability of such a fire and the probability of early detection from smoke. In some instances, materials such as punk may be utilized for its delaying effect, but here again the disadvantages outweight the advantages; such uses are more likely to be read about than encountered. In addition, the primary ignition source remains the flame.

Cigarettes. Cigarettes as a source of ignition for the arsonist will be very uncommon in a successful fire. If they are used to ignite liquid flammables, they will almost certainly fail, as developed elsewhere in this volume. If, more probably, they are used to initiate a smouldering fire, the results will be highly uncertain. Prank fires from this source are more likely than a planned and deliberate arson, because the prankster may throw a lighted cigarette into appropriate fuel, not so much as a premeditated act as simply a caprice. The lighted cigarette in the waste basket is a common source of minor accidental fires from carelessness, and it could result from a prank. While the cigarette alone is a poor source of ignition of a large fire, there are methods of enhancing its value. Instances are encountered in which a cigarette is used in conjunction with matches to produce significant flame. If the lighted cigarette is placed between two layers of paper matches in a matchbook, when it burns down to the heads it will ignite them and the entire book will burst into flame. Another variation is to tie matches around the outside of a cigarette which, when lighted, has the same effect. It is sometimes possible to retrieve the ashes of such a device after the fire to determine that this arrangement was used to set it.

Sparks. Considered as a tool of the arsonist, the spark can be applied only to very restricted types of fuel. Most such fuels are explosive in character, consisting either of explosive gas-air mixtures or of fixed explosives. A fire kindled by means such as these is necessarily characterized by an initial explosion. The only alternative to using the explosive material as initial fuel is to convert the spark in some manner to a flame. This may be done by means of powder trails or, in some instances, by readily ignitable solids such as ordinary or impregnated paper. The use of solids in conjunction with sparks is uncertain and not an efficient method of kindling fires.

A simple exception to the above generalizations is the ordinary pocket lighter in which sparks are utilized to kindle fumes from liquid fuel so as to form a flame before vapor concentrations can build to explosive dimensions. Here, the arsonist must generally be present, and it is the flame that he generates rather than the spark that kindles it that is important.

Because sparks can readily be generated by electrical means which are subject to timed delay, any method of producing a fire by this means would appear to have an inherent advantage to the arsonist. However, without elaborate arrangements of vapors, powder trails, or similar active fuels, and the corresponding elaborate electrical devices which will partially survive the fire, it is difficult to kindle a fire by such means. Further, the evidence that survives may well be traced to the arsonist; this fact, from his standpoint, makes the method undesirable.

Some ingenious methods have been employed, especially by professional arsonists, utilizing sparks that generate when a light switch is turned on or machinery is started. For example, a shortcircuit can be set up by means of just one or two wires of a multiwire cable and this placed in conjunction with readily ignited fuel. Since the shortcircuit is minor, it is difficult to trace, but the heat generated with such a small spark can start a fire. It is also an uncertain method. The spark will continue to be more serviceable for the explosion of bombs than for the production of structural fires and will rarely be encountered.

Glowing wire. Although the author has never encountered an arson case in which the fire was set by means of a glowing wire, it is evident that a timing mechanism such as a clock could be rigged to turn a current through a wire to heat it and ignite fuel that is against the wire. Such a device would allow the fire to be set at any desired time but would be difficult to arrange for rapid build-up. An exceptional device that is claimed to operate excellently involves drilling a hole in a light bulb, filling the bulb with flammable liquid, and screwing it back in place. When the light is turned on, the hot wire both vaporizes and ignites the liquid which starts the fire. Because arsonists are often very ingenious, the investigator must develop both a comparable ingenuity in investigating his cases and a highly developed observational ability.

Chemical Ignition. Various means of kindling a fire by chemical means exist. Most of these are unlikely to be known to any but the professional arsonist and, therefore, will only occasionally be found in arson cases. One of the better known is "thermit," a mixture of aluminum and iron oxide. When ignited, the mixture produces very intense local heat which arises as a result of the difference in the heat of combustion of iron and aluminum, it being much greater for aluminum. The heat produced is sufficient to melt its way through many metals, including iron, and to ignite any combustible in its path. Inevitably, it leaves behind a residue of aluminum oxide and either iron or its oxide, which can be used to indicate the nature of the material. Further, the mixture requires ignition by a primary ignitor, so it serves more to intensify the local combustion than as a source of ignition.

A time-delay mechanism involves the use of acid in a metallic container, which will be eaten through at a slow rate. The acid can then spill into a chemical mixture that will start to burn and produce great heat. Aside from its feature of delaying the combustion, it has little to recommend its use for arson. It is occasionally used in bombs.

Some chemical systems are susceptible to spontaneous ignition under proper circumstances, and means are known for restraining the ignition for a period of time. Among these, hydrogen peroxide and phosphorus are perhaps the most likely to be encountered. Methods of utilizing such materials are known to some professional arsonists and could be developed by a competent chemist, so that there is little doubt that the fire investigator may encounter one or a few such instances in his work. If there is reason to believe that a fire is arson and that any such method of kindling it has been used, the investigator is wise to enlist the aid of a very competent chemist in unraveling the method utilized.

There are a number of additional chemical combinations that can create either explosions or fires, and often on a delayed basis. However, they are not likely to be encountered in arson investigation, because of their inaccessibility to the arsonist or his lack of knowledge of them. When any such unusual means of kindling a fire is present, its detection by chemical means is ordinarily possible and sometimes rather simple. Most chemical systems of this type leave definite residues that can be dealt with by the chemist.

Burning glass. Of little practical interest but some theoretical interest is the lens or burning glass as a source of ignition in arson. Such a glass placed so that the sun would strike it directly at a particular time could serve for approximate timing of a fire set by it. It would also have to be properly focussed in advance and supplied with sufficient fuel material to allow a high probability of ignition and buildup. The uncertainty of the method would argue against its practical use.

Reference

(1) Parker, B. P., Rajeswaren, P., and Kirk, P. L. "Identification of Fire Accelerants by Vapor Phase Chromatogarphy." *Microchem. J.*, 6 31-36, 1962.

14

The Legal Aspect of Arson*

A brief discussion of the background of the law regarding arson and relevant citations and references are included in this treatise for two reasons: (1) the investigator who is familiar with the law governing the area of his investigation has an advantage and (2) there are numerous attorneys who are inexperienced in the matter of arson law. To both, this abbreviated discussion may be useful from time to time.

Criminal burning has always been treated as a serious offense. The old Roman law of "incendium," though broader in scope than common law arson, included willful burning which endangered another's property, setting fire to cities, and causing conflagrations by riot-

*The author is indebted to Mrs. Dorothy H. Northey for the material presented in this chapter.

ing. The penalty for this crime at one time required that the offender be burned alive. As another example, Massachusetts in 1652 enacted a statute which required, upon conviction for arson, death and forfeiture of goods and lands to the party "damnified." That law remained in force until 1784 when the death penalty was only required for arson at night. Arson in the daytime carried a lesser penalty—to be placed in a pillory, whipped, fined, or bound over to good behavior according to the seriousness of the offense.[8] *

Today we find, depending on the jurisdiction, degrees of arson carrying penalties of up to twenty years in prison or even life imprisonment if a death should occur as a result of the criminal act. Some states may have a more severe penalty for burning at night[1] or if someone is in the building at the time.[2]

Now we think of arson as a crime against property, but it was not always considered as such. Common law defined arson as: the willful and malicious burning of the dwelling house of another (either by night or day). This made arson a crime against the security of habitation.[3] There are four important elements of this definition to be considered:

1. *Type of structure.* The structure burned at common law had to be a dwelling house or other structure within the common enclosure. Other buildings were not properly the subject of arson.
2. *House of another.* The structure was required to be the dwelling house of another, with occupancy, not ownership, being the test.
    (a) Thus, to burn one's own house willfully and maliciously was not arson unless someone else lived in a house nearby which was close enough to catch fire. If such were the case, then the requirements of the felony were satisfied.
    (b) Since husband and wife are considered as one at common law, it was not arson for either to burn the home.
    (c) Burning one's own house to defraud the insurer was also not considered as arson.[4]
    (d) A house built, but not yet occupied, was not the subject of arson, even if willfully and maliciously burned. Actual occupancy, rather than intended occupancy, was the test.
3. *Burning.* Actual burning and not mere scorching was (and still is) the test.[5]
4. *Intent.* Willful and malicious are used in the sense of describing a deliberate or intentional act which is done without justification. But the intent may also be inferred where the burning is the natural or probable result of another unlawful act. Burning due to negligence was only trespass at common law and not arson.

*Superscript numbers refer to the citations listed at the end of this chapter.

As the common law concept of arson became inadequate, statutory law expanded the definition to include other buildings and omitted the occupancy test, thus making arson the crime against property as we know it today. Shops, prisons, boats, cars, and public buildings are now included in some arson statutes.

By the 1950's most of the states had adopted The Model Arson Law or had statutes approaching it. This law brought about some measure of uniformity and recognized degrees of arson. The most serious of these is in keeping with the common law concept of arson in that it pertains to dwellings, but all are to be construed in the light of the common law and do not dispense with intent or actual burning requirements. It should be observed in The Model Arson Law which follows that this law also includes principals and accessories and, furthermore, that attempts are defined. It also deals with burning to defraud the insurer, making this a felony but not arson.

California law at the present time does not call burning to defraud the insurer "arson," nor does the pertinent section include the word "malice." It is interesting to note that it is allowable to show evidence of prior insurance collections from previous fires, even though fraud cannot be shown in connection with them, to show possible motive—that the accused was familiar with this source of funds. The reader is urged to read articles by Bolton (13, 14) for the historical development and criticism of arson law in California.

THE MODEL ARSON LAW

Arson: First Degree

Burning of dwellings. Any person who willfully and maliciously sets fire to or burns or causes to be burned or who aids, counsels, or procures the burning of any dwelling house, whether occupied, unoccupied or vacant, or any kitchen, shop, barn, stable, or other outhouse that is parcel thereof, or belongs to or adjoining thereto, whether the property of himself or of another, shall be guilty of Arson in the first degree, and upon conviction thereof be sentenced to the penitentiary for not less than two nor more than twenty years.

Arson: Second Degree

Burning of buildings, etc., other than dwellings. Any person who willfully and maliciously sets fire to or burns or causes to be burned, or who aids, counsels or procures the burning of any building or structure of whatsoever class or character, whether the property of himself or of another, not included or described in the preceding section, shall be guilty of Arson in the second degree, and upon conviction thereof, be sentenced to the penitentiary for not less than one or more than ten years.

Arson: Third Degree

Burning of other property. Any person who willfully and maliciously sets fire to or burns or causes to be burned or who aids, counsels or procures the burning of any personal property of whatsoever class or character (such property being of the value of twenty-five dollars and the property of another person) shall be guilty of Arson in the third degree and upon conviction thereof, be sentenced to the penitentiary for not less than one nor more than three years.

Arson: Fourth Degree

Attempt to burn buildings or property. (a) Any person who willfully and maliciously attempts to set fire to or attempts to burn or aid, counsel or procure the burning of any of the buildings or property mentioned in the foregoing sections, or who commits any act preliminary thereto, or in furtherance thereof, shall be guilty of Arson in the fourth degree and upon conviction thereof be sentenced to the penitentiary for not less than one nor more than two years or fined not to exceed one thousand dollars.

Definition of an attempt to burn. (b) The placing or distributing of any flammable, explosive or combustible material or substance, or any device in any building or property mentioned in the foregoing sections in an arrangement or preparation with intent to eventually willfully and maliciously set fire to or burn same, or to procure the setting fire to or burning of same shall, for the purpose of this act constitute an attempt to burn such building or property.

Burning to defraud insurer. Any person who willfully and with intent to injure or defraud the insurer sets fire to or burns or attempts to do so or who causes to be burned or who aids, counsels or procures the burning of any building, structure or personal property, of whatsoever class or character, whether the property of himself or of another, which shall at the time be insured by any person, company or corporation against loss or damage by fire, shall be guilty of a felony and upon conviction thereof, be sentenced to the penitentiary for not less than one nor more than five years.

Every fire must be presumed to be accidental until or unless evidence of the corpus delicti of arson can overcome this presumption. The corpus delicti of arson requires more than mere proof of burning.[6] It also requires that the burning be the result of a criminal agency.

To satisfy the burning requirement, it is not necessary that a whole building be destroyed. As long as any fiber thereof is destroyed or altered in identity it is sufficient. See Peo v Haggerty, 46 Cal 354, 355

"To 'burn' means to consume by fire and in a case of arson, if the wood is blackened, but no fibers are wasted, there is no burning, yet the wood need not be in a blaze, and the burning of any part, however small, is sufficient to constitute the offence, and if the house is charred in a single place so as to destroy any of the fibers of the wood, it is sufficient to constitute arson."

Nor is it necessary for the fire to continue for any length of time.[7] In fact, it may go out by itself and still be arson. In Jones v Comm, 271 Ky 647, 113 SW 2d 7, it was held proper for the judge to instruct the jury that it is arson if there is burning to "any extent."

Statutes may read "burn" or "set fire to" and in general these are held to be synonymous. Arson usually refers to fire. However, in some jurisdictions, if in an explosion, wood or other combustible material is damaged, it is within the law.[8] If indirect means are used to convey fire from one place to another it is still arson if there is intent to burn both.

Also, part of the corpus delicti to be proved is that the burning is intentional rather than accidental. As noted before, the natural and probable consequences of a person's acts are presumed to be the intended consequences. The word willful as used in the statutes means intentional or with knowledge and purpose.[9] When a person unintentionally sets fire to a building in the course of committing another felony, the felonious intent is also attributed to the arson.[10]

Where several origins are shown along with evidence showing the improbability of accident, it is considered sufficient evidence of the corpus delicti.[11,12,13]

While intent is an element of the crime of arson, motive is not. However, motive may be relevant to help identify the accused as the responsible party, but it is not evidence of the crime itself.[14]

Confessions, as in other criminal cases, may in some cases be unacceptable unless the corpus delicti can be established. In other words, the confession must be corroborated by other evidence.

In addition to proving the corpus delicti, the prosecution must prove that the defendant is the responsible criminal party. Both factors must be proved *beyond a reasonable doubt.* As in Peo v Andrews (1965), 234 Cal App 2d 69, 44 Cal Reptr 94, corpus delicti may be established even though the showing of incendiary origin stops short of absolute certainty.

Arson may be proved by direct or circumstantial evidence. However, it is not often that there is an eye-witness to the crime. Most often the evidence is largely circumstantial. This means that the crime and guilt of the accused must be inferred from other facts which can be established. There are limitations on conclusions which are based on inferences. That is, the incriminating facts must be consistent with guilt of the accused and inconsistent with any other rational conclusion.[15, 16]

There are differences with respect to the admissibility of testimony given by eye-witnesses and experts. In general, experts are allowed to express opinions and eye-witnesses may only testify to facts observed. In some cases, testimony as to the cause of a fire by an eye-witness may be acceptable as a statement of fact where the testimony does not extend past the facts observed. Likewise, in some cases, opinions by the expert are excluded where the opinion is given as to the ultimate cause or origin of the fire (holding this to be an issue for the jury). Perhaps the better opinion and the modern trend is expressed in two cases upholding the expert testimony. Comm v Nasuti (1956), 385 Pa 436, 123 A2d 435, held that the ultimate question was the guilt of the defendant, not the incendiary origin. Also, Comm v Kaufman (1956), 182 Pa Super 197, 126 A2d 758, held such evidence does not invade the right of the jury because the jury does not have to accept the expert opinion where the facts indicate another view.

Anyone with special knowledge or experience with respect to the subject in question may qualify through a showing of training or experience to testify as an expert. In the case of arson, the expert giving opinion testimony as to the cause or origin of a fire may be a fire investigator, criminalist, or one of any number of specialists—chemist, electrician, etc., depending on the nature of the fire. Testimony by such an individual stating that the fire is of "incendiary origin" or "set" is admissible where it is determined that the cause or origin is not within the common knowledge of the layman. Opinion testimony tending to negate incendiary origin is admissible from experts but not from the ordinary witness.

Normally the weight of evidence as to the cause of a fire is determined by the same tests applied to other evidence. In other words, the jury is not bound to accept it. Also, experts may be allowed to introduce evidence of experiments to sustain a theory of how the fire started if the conditions surrounding the experiment are not too dissimilar to the actual event.[17]

Usually, incendiary origin of fire is established by circumstantial evidence showing that a highly flammable liquid, or its odor or container, was present, or by showing the improbability of an accidental fire—for example, three separate origins.[18]

As with any other crime, there are limits of responsibility or defenses. If the accused meets the tests for "legal insanity" for a given jurisdiction then he is insane by law and therefore not responsible. Infancy is also *prima facie* evidence of incapability to commit arson. But voluntary intoxication is no defense, nor is consent of the owner a defense. If circumstances would make it a crime for the owner to commit the act, then it is also arson on the part of the individual performing the act for the owner. Where burning is for the purpose of defrauding the insurer, it is no defense to show that the policy was not in force at the time of the burning. It is sufficient if the accused believed the policy was in force at the time the act was committed.

Citations

[1] State v Haynes, 66 Mi 307
[2] State v Lockwood, 24 Del 28, 74 A2
[3] State v McGowan, 20 Conn 245
[4] Roberts v State, 47 Tenn 359
[5] Dedieu v Peo, 22 NY 178 (not sufficient if near or against, must burn)
[6] Peo v Holman, 72 CA 2d 75, 164 p2d 297
[7] Mary v State, 24 Ark 44
[8] State v Murphy, 214 La 600, 38 So 2d 254
[9] Peo v Andrews, 234 Cal App 2d 69, 44 Cal Reptr 94
[10] Peo v Fanshawe, 137 NY 68, 32 NE 1102
[11] Peo v Sainder, 13 CA 743 110 p825
[12] Peo v Patello, 125 CA 480, 13 p2d 1068
[13] Peo v Hays, 101 CA 2d 305, 225 p2d 600
[14] Pointer v U.S., 151 U.S. 396, 413, 38 L Ed. 208, 216
[15] Peo v Lepkojes, 48 CA 654 192 p 160
[16] Peo v Kessler, 62 CA 2d 817, 45 p2d 656
[17] Peo v Sherman, 97 Cal App 2d 245, 217 p2d 715
[18] Peo v Saunders, 13 Cal App 743 110 p 825

References

(1) Burdick, W. L. *Law of Crimes* (Vol. 3) Matthew Bender, 1946.
(2) Clark and Marshall. *A Treatise on the Law of Crimes*, 7th Ed., Barnes, M. Q., Revising Ed., Callaghan & Co., 1967.
(3) Braun, W. C. "Legal Aspects of Arson." *J. Crim. Law, Criminol. and Police Sci.*, **43**, 53, 1952.
(4) Cohen, H. H. "Convicting the Arsonist." *J. Crim. Law, Criminol. and Police Sci.*, **38**, 286, 1948.
(5) Hopper, W. H. "Arson's Corpus Delicti." *J. Crim. Law, Criminol. and Police Sci.*, **47**, 118, 1956.
(6) Stevens, S. L. "Evidence of Arson and Its Legal Aspects." *J. Crim. Law, Criminol. and Police Sci.*, **44**, 817, 1953.
(7) Battle, B. P. and Weston, P. B. *Arson: Handbook of Detection and Investigation*, Greenberg, N. Y., 1954.
(8) Rethoret, H. *Fire Investigations*. Recording & Statistical Corp., Limited, Canada, 1945. See p. 25.

(9) 5 Cal Jur 2d, Arson and Willful Burning.
(10) *New Calif. Digest*, McKinney 3A, 1963.
(11) 88 ALR 2d 230 (*Am. Law Reports*).
(12) *So. Cal. Law Review*, **XXII** (3), Apr. 1949.
(13) *So. Cal. Law Review*, **XXXV**, 375, 1962.
(14) Am. Juris. 2d, 1962.

15

Carbon Monoxide Asphyxiation

Although asphyxiation and fire are separate types of events, virtually all asphyxiations result from a fire in which the combustion is incomplete. Since, for practical purposes, only carbon compounds (organic compounds) burn, it can be considered that the ultimate combustion product of such burning is carbon dioxide. However, the production of carbon dioxide is related chiefly to the degree of ventilation of the fire, such as the available oxygen, and to the temperature(1).

In nearly all fires, the supply of oxygen of the air is insufficient to allow complete combustion of all the carbon compounds. Thus, carbon monoxide is an almost constant factor in all fires. In structural fires especially, it is the rule that if anyone is killed, it is nearly always because he inhaled the gaseous products of fire.

It is rare in such instances for persons to be burned until after they had died from carbon monoxide inhalation. Usually, these are persons who are sleeping on an upper floor above a fire burning on a lower floor. In the end, their bodies may be charred, but the cause of death is determined to be carbon monoxide, and the blood analysis is likely to show 70 or 80 or even 90 percent saturation with this gas. In the investigation of the fire, data such as these can assist with the analysis of the course of the fire which was the cause of their deaths. It is thus important to analyze the details of carbon monoxide asphyxiation.

NATURE OF CARBON MONOXIDE ASPHYXIATION

The classification of deaths from carbon monoxide has always been troublesome. The gas itself is not toxic in the usual sense of interfering with normal metabolism directly, yet it is probably the first (2) or second most important cause of death from "toxic" materials. Carbon monoxide, like oxygen of the air, will combine with hemoglobin of the blood. The difference is that carbon monoxide combines with about two hundred times the avidity or stability of the complex, as compared with oxygen (3). Since it is the blood hemoglobin that transports all the oxygen that reaches the tissues, and therefore is a critical link in the chain that supports normal oxidative metabolism and life itself, the intrusion of a more stable material that displaces the oxygen with a stable and useless hemoglobin combination simply serves to starve the cells of needed oxygen, and an internal suffocation results. As a corollary to this effect, a person with a given quantity of carboxyhemoglobin (hemoglobin combined with carbon monoxide), when exposed to an atmosphere containing more than two hundred times as much oxygen as carbon monoxide, will allow the oxygen to displace the carbon monoxide from the hemoglobin, and the content of the latter will diminish proportionately. Thus, if a person after exposure to carbon monoxide in a dangerous concentration is placed in clean air, or better, given oxygen (4), he will generally recover as a result of this displacement.

Based on relative stability of carboxyhemoglobin and oxyhemoglobin, it is relatively simple to calculate the carbon monoxide content of the air which may be dangerous. If the concentration of oxygen in air is just two hundred times as high as the carbon monoxide, extended breathing of this mixture would result in forming a mixture of carboxyhemoglobin and oxyhemoglobin which would be about equal to each other, and one-half of the effective oxygen carriage of the blood would be lost. Since air contains about twenty percent of oxygen, one two-hundredth of this value would be 0.1 percent. Thus, breathing of a concentration of 0.1 percent or 1000 parts per million would rather rapidly become fatal, since removal of one-half of the oxygen-carrying capacity of blood is not compatible with continued life. Experience shows that the dangerous limits are actually about 0.02 percent, and that over a concentration of carbon monoxide of 0.06 percent, the air can be breathed only for a period of one and one-half hours without becoming critically dangerous. At concentrations below this value, down to 0.01 percent, symptoms such as mental disorientation, slow reaction time, and other conditions that are not compatible with operation of machinery or vehicles may ensue after two to four hours of exposure (5). Values in this range are actually reached as a result of severe air pollution, especially when the person breathing such a mixture is working in some type of enclosure in which the ventilation is restricted.

THE CARBON MONOXIDE HAZARD

Everyone is exposed to some concentration of carbon monoxide in the air he breathes, because there is measurable amount of the gas in nearly all of the atmosphere. In urban areas especially, the concentration becomes relatively high, especially because of the automobile exhaust which is always rather rich in carbon monoxide. Smoking produces a comparatively high concentration of carbon monoxide in air that reaches the lungs of the smoker. From these causes alone, it is not uncommon to find a smoker with as much as five percent of his blood hemoglobin combined with this gas. Rather high figures also may be obtained from those who breathe considerable amounts of automobile exhaust. A very certain method of committing suicide that has long been known is to connect the interior of a car, by a hose, with the exhaust pipe. It must not be forgotten that this carbon monoxide is also the product of fire, even though that fire occurs within the cylinders of an automobile engine.

Regardless of the exposure of the ordinary citizen to carbon monoxide in quantities that show distinctly in his blood, such exposure is not expected to result in fatalities. Although a five percent saturation level of carboxyhemoglobin may be shown in many ordinary citizens, it requires about twenty percent saturation to produce serious symptoms such as a headache, nausea, disorientation, and dizziness. At about forty percent saturation, unconsciousness is expected, although the results may be more or less extreme. Bodies, determined to have been killed by carbon monoxide asphyxiation, have been found to have levels close to and below forty percent; however, this is unusual. It is generally agreed that a level of sixty percent saturation or greater is incompatible with life, despite the finding in dead bodies of both lower and higher degrees of saturation. Data in Table 1, quoted from Gettler and Freimuth (6) as given by Stolman and Stewart (4), indicate the carboxyhemoglobin percentage in a total of sixty-five cases of fatal asphyxiation. It will be noted that they found some deaths below forty percent and none over ninety percent. The author has encountered laboratory values of over ninety percent in a few instances.

TABLE 1. FATAL CASES OF CARBON MONOXIDE
ASPHYXIATION CLASSIFIED ACCORDING
TO CARBON MONOXIDE SATURATION
OF THE BLOOD

Total cases (%)*	CO (% saturation of blood)
4.5	30-40
4.5	40-50
14.5	50-60
28.0	60-70
48.5	70-80

*Total of sixty-five cases.

EFFECT OF RATE OF ABSORPTION

It is well recognized that all individuals show differences in tolerance to the influence of toxic materials and to other basically environmental influences. Certainly, such variations must exist with carbon monoxide as well. For example, one person may have a considerably higher content of hemoglobin initially than another. In such an instance, it would be expected that the first would require a greater loss of hemoglobin before a fatality would result than would be true of the second. However, it is not realistic to believe that any person would be capable of continued survival with half of his hemoglobin inactivated by its combination with carbon monoxide. If this assumption is a reasonable approximation to the truth, it remains to explain how one person who dies from cabron monoxide at forty percent saturation can be consistent with another who accumulated a ninety percent saturation before death. The factors appear to be as follows:

1. Rate of inhalation of the gas.
2. Activity or its lack which changes requirements for oxygen.
3. Individual variations in susceptibility.

The first item is probably far more important than the others. Remember that death from oxygen starvation is not a rapid process; persons who have already inhaled a lethal quantity of the gas will continue to build the concentration of carboxyhemoglobin past the value that is inevitably lethal. Consider two cases. In the first, the person is exposed to a marginally excessive quantity of carbon monoxide. His blood saturation increases slowly, maintaining essential equilibrium with the gas in the air he is breathing. When enough of his blood hemoglobin has been effectively destroyed, he will die, but the carboxyhemoglobin concentration will be minimal.

In the second instance, assume that the person suddenly is exposed to a very large increase in the carbon monoxide content of the air which carries it greatly over the minimum danger level. The time lag in absorption and in the final fatal results, allows him to breathe high concentrations over a period of time, even after he has passed the point of no return. These are the persons whose blood may show eighty or ninety percent saturation of the hemoglobin with carbon monoxide.

The significance of this consideration in the investigation of a fire is evident. If the dead body shows only a minimal saturation compatible with death, it indicates a long exposure to a relatively low concentration of gas, which might correspond to a small, possibly smouldering fire that continued for a considerable time. On the other hand, if his blood shows a very high percentage saturation, it would indicate a much shorter exposure to a high concentration, which would be consistent with a much greater fire that suddenly increased the content of carbon monoxide to a relatively high value. Such information can usefully supplement that which is obtained by examination of the fire scene itself.

Item No. 2 serves to modify the rather sweeping generalities of the above paragraphs. If a person is relatively inactive, his need for oxygen is diminished, his breathing is slower and more shallow, and carbon monoxide in the air will affect him much less than if he breathes deeply and at a high rate, as would be required in performing physical work. Thus, it can be stated that if a person suspects that carbon monoxide is present in significant amounts in the ambient air, he is better off if he rests quietly, breathes as little and as shallowly as possible, and avoids all activity that will increase his breathing rate. Such considerations are often important for firemen who must exert themselves in fighting a fire, rescuing trapped persons, and the like, because the high degree of activity will inevitably increase considerably the hazard to themselves.

Item No. 3 is of less importance in connection with carbon monoxide inhalation than with ordinary drug or poison reactions. It has significance, but largely in the area of physiological differences, concentration of hemoglobin of the blood, and related factors that are generally not under control by the individual. It is not expected that such differences would be as pronounced with carbon monoxide as with most physiologically active agents, because of the rather unique type of effect of the gas in interfering with a normal process through an easily measurable mechanism.

SOURCES OF CARBON MONOXIDE

It has been stated earlier that virtually all carbon monoxide is the product of incomplete combustion of carbon compounds. To be more specific, almost every instance of asphyxiation from this gas results from one of the following:

1. Defective *heating equipment*, probably the primary source of fatalities from carbon monoxide asphyxiation.
2. *Motor exhaust*, which invariably contains dangerous amounts of carbon monoxide, even with a well-adjusted carburetor.
3. *Industrial processes*, in which carbon monoxide either results from unintentional incomplete combustion processes that are part of the operation or from deliberate formation of carbon monoxide uitilized as a portion of the industrial process.
4. *Fires involving structures* and occasionally exterior fires.

Heating Equipment

Heating equipment is always suspect when a body of an asphyxiated person is found in his residence with no obvious connection to automobile exhaust. Some general principles of the operation of heating equipment (other than electrical) are important to understand. In order to create an environment that is dangerous because of its carbon monoxide content, two things are essential: (1) the fire maintained in the heating equipment is generating carbon monoxide in dangerous quantities, and (2) the gas so generated is escaping into the room air. *These two points are so important that they must never be overlooked.* Most flames generate some of the gas, and any yellow or orange flame is probably generating a great deal of it. A blue flame from gas, for example, is generating very little carbon monoxide and would not be hazardous ordinarily, even if all of its effluent gases were escaping directly into the room. It can be stated that venting of gas appliances is actually not really necessary if the flame were at all times properly adjusted. Since flames do not always stay adjusted, there remains a strong and valid argument for the presence of adequate venting which is required by the codes of most municipalities.

Proper performance of a gas flame depends on mixing the gas with air in exactly the correct proportion. When the mixture is of correct composition, there is little or no carbon monoxide formed. The initial mixing is performed by the venturi, and adjustment of the flame is done completely by its manipulation. The venturi consists of an internal nozzle through which a relatively high velocity but small stream of gas is forced by the pressure in the gas line. Surrounding this nozzle is a larger chamber that is open to the air. The jet of gas is cone shaped and should fill the throat of the outer chamber, drawing air in through the opening. Secondary adjustment of the available air is provided by some sort of shutter mechanism which can be opened and closed for final adjustment. This mixture of gas with air is the combustion mixture which passes to the burner, where still more air becomes available to the flame (secondary air supply).

There are a number of things that can affect the operation of the venturi in producing a proper mixture, including the following:

1. *Barometric pressure* changes, affected both by *altitude* and by fluctuation of the ambient air pressure.
2. *Size of orifice* in the nozzle, which by being too large can provide more gas than the venturi can properly mix with air, or by being too small will give an unstable flame.
3. *Improper drilling* of the orifice in the venturi nozzle, e.g., at some angle which does not direct the cone of gas directly at the throat and produces turbulence and poor drawing of air into the stream.
4. *Irregularities in the inner surface of the throat* of the venturi, which produce a turbulent gas flow with reduced air admixture.
5. *Obstructions* such as balls of lint that tend to suck into the venturi throat and build up, so as to impede the gas velocity and produce dangerous flames.
6. *Improper adjustment of the air intake* so that even a properly assembled venturi cannot draw enough air.

Insofar as installation and initial adjustment of the appliance is concerned, the size of the nozzle orifice is probably most important. All gas appliances are designed for a particular BTU output, which translated into simple terms means a particular rate of furnishing gas to the flame, and affecting accordingly the amount of heat produced. The higher the BTU value, the more gas is consumed. But if the nozzle is drilled for a particular value when the remainder of the appliance is not designed for so high a value, there is a tendency for flame actually to spill from the appliance, and the air supplied is definitely inadequate for total combustion. Such an arrangement invariably produces excessive amounts of carbon monoxide.

If the size of the orifice is correct for the particular appliance design, but other factors listed above are operative, the flame will still become yellow, form soot, and, especially, generate carbon monoxide in excessive quantity. These are the items that must be checked with any gas appliance that has produced an asphyxiation, and it is not ordinarily a difficult task to locate the source of the difficulty. Almost invariably it resides in some improper adjustment or condition of the venturi. Some of these effects are illustrated in Figure 3.

Having established the cause of generation of excessive amounts of carbon monoxide, it is then necessary to locate the reason that the venting of the appliance did not harmlessly remove the gas but rather delivered it into the air that was breathed by the victim of asphyxiation. This always has to do with the system by which combustion gases are vented or kept from entering the environment of the persons utilizing the appliance.

Escape of burner fumes into the ambient air is the result of some defect in the heating unit itself or in the vents that are normally present. In one instance, a dead body was present in the bedroom. The apartment heater had been

188   *Fire Investigation*

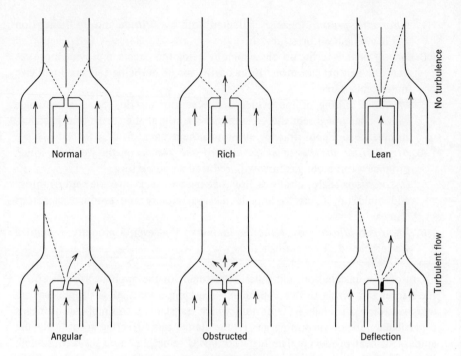

Figure 3. *Variations in Venturi operation.*

separated from its vent pipe, leaving a gap of about one inch. When the hot gases reached this gap, they rose into the room air rather than continuing into the vent system. In another system, a furnace, there were open gaps in the fire box which served then not only as a heat exchanger with the air circulation system but also contributed the burner gases directly to that circulating system, resulting in two deaths. In still another, the living quarters were so tightly sealed against cold air that a partial vacuum was created in the room, thus causing diversion of the burner gases through the draft diverter back into the room. One death and two injuries resulted.

When heating equipment is generating carbon monoxide, there will be found in nearly every instance two conditions symptomatic of the situation: (1) the color of the flame, at least at the top, will not be blue, but orange or reddish, and (2) there will be soot deposits above the flame, on heat exchanger boxes, on vents, and frequently even on the outside of the unit. When these symptoms are located, it is certain that dangerous quantities of carbon monoxide are being generated.

Automobile Exhaust

The danger from carbon monoxide of automobile exhaust takes two directions:

(1) direct asphyxiation, which is commonly of a suicidal nature, although it may also be accidental, and (2) air pollution in regions where the density of automobile traffic is excessive as in some metropolitan areas. It is clear that the situations in the two instances are quite different, and that both constitute definite hazards.

Direct asphyxiation will occur when the environment is sufficiently restricted in volume that the exhaust output can produce a significant and dangerous concentration of carbon monoxide. Running a motor in a closed private garage is a common source of dangerous and sometimes fatal concentrations. Damaged exhaust systems, which include those with a loose exhaust manifold, corroded exhaust pipes and mufflers, and similar leaky conducting systems for the exhaust gas, may provide a concentration of the gas locally in such a location that it can penetrate the interior of the vehicle. Under these circumstances, the interior concentration builds up correspondingly and may reach lethal concentrations. This also is likely to be entirely accidental. In cases of suicide, it is more common for the exhaust pipe to be connected to the interior of the vehicle with a hose, so that the entire effluent gas is delivered inside. The presence of such a hose connected with the exhaust pipe is, of course, a positive indication of deliberate arrangement and suicide.

The more insidious danger of carbon monoxide from the automobile exhaust occurs when motor repair and adjustment involve operating the motor in a small and relatively well-enclosed space, as in cold weather when excessive ventilation is avoided because of the cold. Even rather large public garages may be subject to this hazard, which may well be neglected or misunderstood.

Airplanes exhibit at times some severe hazards from carbon monoxide. Military craft and small private planes appear to be the only ones in which this difficulty has been noted. The origin of the gas in these craft is not always simple to trace, possibly because of the high velocities of gas streams associated with air-borne vehicles. At least in some instances the problem appears to have originated from the necessity of considerable heat being delivered to the cabin. Since the motors are air cooled, there is no opportunity to attach heaters to a water-cooling system, as in an automobile, and exhaust heat has been used. If there is any leakage of exhaust gas into the heater as a result of corrosion or loose connections, the occupants of the cabin may be suffocated very rapidly. It is believed that this situation has been the cause of a considerable number of crashes of light planes.

Air pollution, the second listed hazard from automobile fumes, is scarcely an appropriate subject for treatment in these pages. It does constitute a present danger, and one with which the public must be concerned increasingly, as has been demonstrated adequately in recent studies. Although unlikely to produce immediately lethal results, it can keep the blood carboxyhemoglobin at a high enough level to cause serious impairment of efficiency and lead to other accidents as a result.

Industrial Processes

In most instances in which carbon monoxide is deliberately generated for industrial purposes, the dangers associated with the operation are recognized and appropriate measures taken to minimize them. However, the ever-present factor of human carelessness can and sometimes does overcome all of the positive safety measures that are in effect and lead to tragic results. In one instance, aluminum powder was being protected from dangerous and undersirable oxidation by keeping it under a blanket of gases with a high content of carbon monoxide. This was led to the tank by means of a hose which was not intended to be opened for gas flow until the vessel was sealed. A workman opened the hose while the vessel was not sealed, effectively blowing the gas directly at his face. The carelessness resulted in a fatality.

Even more dangerous is the combustion process occurring in the industrial plant, in which carbon monoxide is generated freely and without sufficient precautions to exhaust it. Lack of appreciation of the danger or simple human carelessness in such an instance may well lead to asphyxiation. The possibilities are so numerous that no detailed discussion of them can be included, but they are very real, and the investigator must be prepared to appraise them in actual instances.

Structural Fires

In every structural fire, large quantities of carbon monoxide are generated and, as stated above, are the cause of most of the fatalities accompanying fires. Similar considerations apply also to the exterior fire, such as a forest fire, but here the ventilation is better and asphyxiations are not likely. In the structure, there is a channeling of effluent gases, and anyone who happens to be in the channel is exposed to the carbon monoxide. Little can be done to minimize this danger other than the general measures that prevent the fire itself, so that the importance of asphyxiations under these circumstances, in addition to their tragic consequences, is in the aid they can provide to the investigator, as discussed earlier.

INVESTIGATION OF CARBON MONOXIDE ASPHYXIATIONS

Determine probable origin. The circumstances of the asphyxiation will normally throw heavy suspicion on a particular source. In a house that has not burned, gas appliances are first choice as offenders. If a fire occurred in the house, that is the most probable source. In a car, it is almost certainly motor effluents. Such probable sources are generally apparent to the first person on the scene, and they require little effort to pick them.

Make sure that the victim actually died from carbon monoxide. The coroner's report will in such cases include a carboxyhemoglobin value. If there was asphyxiation, the value will generally be over forty percent but may occasionally be a little lower. If much lower, there will be real doubt as to the cause of death, and the investigation may take a different direction.

With structural fires. Investigation can generally stop at this point, because an obvious cause was present, and there is no remedial action to take.

With automobile asphyxiations:
1. Check for possible suicide by searching for artificial means, such as a hose, by which gas was introduced into the car, seeing if windows and vents are closed or open, and checking the setting of throttle. If findings are negative, proceed to No. 2.
2. Examine exhaust system closely from manifold through to tailpipe. At some point of junction of the parts, there may be a sooty area fanning from a seam. This is especially significant if it is in the forward rather than the rear portion of the vehicle. Exhaust manifolds may require special attention because soot may burn off of them, but actual gaps may be located. Having found such a point of escape from the exhaust system, proceed to No. 3.
3. Determine the manner by which the gas entered the interior of the vehicle. Holes in the firewall, combined with leakage around the manifold, will produce contamination of the interior air by gases blown against the firewall by the motor's fan. If leakage is from a muffler, the condition of the floor is critical, because there has to be a means of passage of gas into the car.

With gas appliances. The *first item* is to test the venturi operation. The flame is lit and observed for yellow color, and if visible in the appliance, the tip of the flame is observed for tell-tale smokey streams. It is assumed that soot is present, and that its location will indicate the defective burner if there is more than one. Having observed a defective flame, observation must be made as to whether it is spilling from the firebox, which indicates a too large orifice in the nozzle. At this point, a carbon monoxide detector, if available, should be utilized to prove the presence and approximate quantity of the gas in the air around the heater and at the vent. The throat of the venturi, as well as the air adjustment, is checked. If the air adjustment is closed too tightly, merely opening it may correct the flame. If open, there is either too much gas being furnished to the venturi or there is interference with the air supply, such as a turbulent condition previously described. Inspection will reveal presence of foreign material in the venturi throat or defects of the interior surface. The venturi may have to be dismantled and further measurements on the orifice taken, or other more detailed observations made.

If the indications are for too much gas without other defects, the gas utility should be contacted to attach a special meter and measure the gas input as compared with the BTU rating of the appliance. Very often, a new and oversized spud has been screwed into the nozzle in order to correct service difficulties, adjust for a change in the gas supply, etc. Sometimes, service personnel will redrill the nozzle to a larger diameter and choose the incorrect drill size. There are other contingencies that may arise and be detected, but it is almost invariably safe to attribute all the deficiencies of a flame to something incorrectly arranged in the venturi.

The *second item* of the investigation of the gas appliance is to locate the reason that carbon monoxide from the flame is escaping into the room rather than being harmlessly vented. Sometimes this results from separation of joints in the firebox or heat exchanger, which can be observed. Sometimes it results from a plugged vent pipe or flue. A reliable indicator of where the difficulty is may often be the region that carries excessive quantities of soot, and this is always the first thing that the investigator should look for. If there is no soot associated with the appliance, it is even probable that some other source of the carbon monoxide must be sought, because soot is an almost invariable accompaniment of dangerous quantities of carbon monoxide.

Having located both the source of the gas generation in large quantity, and the manner in which it escapes into the room air rather than being vented to the exterior, the investigation is essentially complete, and a total and accurate description of the difficulty can be given.

References

(1) Golovina, E.. S. and Khaustovich, G. P. 8th Symposium (International) on Combustion, California, Aug. 28-Sept. 3, 1960, p. 784.
(2) Thienes, C. H. and Haley, T. J., *Clinical Toxicology* Lea & Febiger, Philadelphia, 1948.
(3) Hawk, P. B., Oser, B. L., and Summerson, W. H. *Practical Physiological Chemistry* (12th Ed.), Blakiston Co., Philadelphia, 1947.
(4) Stolman, A. and Stewart, C. P. *Toxicology, Mechanisms and Analytical Methods* (Vol. 1). Academic Press, New York, 1960.
(5) Technical Paper 212, Natl. Bureau of Standards, Washington, D.C.
(6) Gettler, A. O. and Freimuth, H. C., "Carbon Monoxide in Blood: A Simple and Rapid Estimation." *Am. J. Clin. Pathol. Tech. Suppl.* 7, 79, 1943.

16

Explosions Associated with Fires

Fires and explosions so frequently accompany each other that no treatise on fire and its investigation would be complete without some special consideration of the nature and role of explosions. The diffuse type, characteristic of gases and vapors, which when admixed with air will produce almost instantaneous combustion with the characteristics of great mechanical force and those manifestations ordinarily called explosions, has been mentioned repeatedly in context. The relative evaluation of such events with the true explosion, which also may produce a fire, requires further elaboration.

There are two types of oxidative combustion that are known as "explosions." The first and most common of these is the type already considered, the *diffuse explosion,* which is actually not a true explosion but a very rapid combustion resulting from having a mixture of fuel in the form of gas, vapor, or fine dust admixed with an appropriate quantity of air to allow almost total combustion to proceed within a very short time.

The second is the *concentrated explosion*, which is not dependent on air but results from a combustion or other chemical reaction in a true explosive. By true explosive is meant either a material that carries not only the fuel for the combustion but an internal oxidant as part of the explosive, or alternatively, it consists of a very unstable material which is capable of rearranging the molecular structure violently when detonated, or heated, and with strong exothermic effects. Dynamite, TNT, propellant powders, nitroglycerine, and many others are illustrations of true explosives which always include in themselves all the essential components for the almost instantaneous reaction of the explosion.

Present concern with the true explosive is limited in this volume because it is not a common source of fire, although it may initiate a fire. On the other hand, the dividing line between true explosives and systems that produce similar results without meeting the requirement of containing all reactants is not sharp and must be considered.

Physically, there are differences between the diffuse and the concentrated explosion, and here also the distinction is not necessarily sharp. The major differences are in the *reaction time* and the *reaction volume* of the explosive reaction. Diffuse explosions, although they appear to be instantaneous, actually require much more time for the total reaction than is the case with the true explosive. This difference may be in orders of magnitude, so that it is not a minor distinction; but to the witness of the explosion, the difference is not likely to be perceptible. The forces generated also are different, but more because of the time relation than because of the total force generated.

A more important physical difference lies in the fact that the concentrated explosive is literally concentrated in space. Assume that the total explosive systems of a stick of dynamite in a room on the one hand and the same room full of explosive vapors on the other were to explode. The *volume* of the first explosion is the volume of a single stick of dynamite, essentially a point source in the room, where all the forces are generated. In the second instance, the explosive volume is that of the entire room. In physical terms, this is the chief distinction between the concentrated and the diffuse explosion. It is evident that the results of the two explosions would be quite different, even though the forces produced might be quite comparable.

DIFFUSE EXPLOSIONS

Perhaps the most common type of diffuse explosion is that created when natural gas escapes in quantity into a more or less confined space and mixes with air in that space. From the combustion limit table of Chapter 4, it is seen that if one volume of the escaping gas is mixed with about ten volumes of air, an explosive mixture results close to the theoretical ratio of gases. If this approximately fills the space and is ignited, the actual explosion is a combustion throughout the entire space at just about the same moment. The combustion produces a great amount of heat suddenly, the combustion gases (and nitrogen of the air) expand violently, and the limiting walls of the space in question are blown outward. In the process, a great deal of heat is generated within a very short time limit. Items that are very readily ignited, charred, or damaged will be affected. Thus, eyebrows may be singed, but flesh not burned, because a longer time for heat penetration is required for flesh than for hair. Clothing will be set on fire, but wooden objects will be little affected, because clothing fibers can kindle rapidly as compared with wooden objects. While the force produced may move the walls outward or blow the roof upward, a person close to the center may actually suffer little effect. If the compression is from all sides simultaneously, he is not moved; if the heat is sufficiently transitory, he is not seriously burned. It is common in such explosions to observe such apparently puzzling discrepancies that are, in actuality, not discrepancies at all when their reasons are understood.

Consider now the "rich" mixture explosion. Suppose that the gas escapes to an extent that the space is filled with a mixture of twelve or more parts per hundred of natural gas in the air, or that there are pockets in which the gas concentration may be much higher even than this. When the mixture is ignited, there is an explosion, but at no point is there sufficient air (oxygen) available for total and immediate combustion. The gas is not totally burned, and the force of the explosion drops off accordingly. However, there is gas left over, and it has been heated far above its ignition temperature. It is expanded when heated, but cools immediately afterward, and air is drawn into the partial vacuum that results. This provides more oxygen to combine with the residual hot gas, and a "rolling fire" results. This particular combination is quite common in practice, and is generally described as, "I heard a 'whoosh' and was on fire." Actually, the destruction by "rich" mixtures may far exceed those by essentially theoretical mixtures of gas and air, because the explosions are followed by fire, and the fire may consume a building that otherwise would have been subjected only to a sudden expansion that might move its walls but not consume its contents.

The "lean" mixture is of less significance than the others when an explosion results. The explosion is sharper, because the propagation rate is greater in the lean than in the rich mixture. Thus, its force may be considerable, its noise very startling, but its damage minor by comparison with the others. No fire is expected from a lean explosive mixture, except perhaps that some filmy material such as a kitchen curtain may be ignited and may possibly set other closeby fuel on fire. Generally the results of an explosion of this type of gas mixture are purely in the area of physical force, not of succeeding fire, but this cannot be guaranteed in every instance.

VAPOR EXPLOSIONS

In the above discussion, natural gas was chosen as the illustrative fuel because it is lighter than air and mixes rather freely with air, thus demonstrating nicely the variations produced by percentage admixture of gas and air. Natural gas is one of only two or three gases that are lighter than air. *All vapors generated by evaporation of liquids at ordinary temperatures are heavier than air.* For example, gasoline vapors are about four times as heavy as air. Most solvent mixtures, naphthas, and similar liquids are even heavier and more non-volatile than gasoline. There are exceptions, which the investigator must consider carefully, but even the exceptions are definitely heavier than air. It must be assumed, in these instances, that the vapors are heavy enough to concentrate at the bottom of any enclosed space and to form there a very "rich" mixture. As illustrated previously, this very rich mixture will be covered by one within the explosive (or flammable) range, and generally this may be covered with a non-flammable mixture, too lean to even explode. When such a condition exists, the explosion will have little force, but the rolling fire may be very destructive. Here the "whoosh" will take over from the "bang," but the structure will probably suffer great destruction from the resulting fire.

All of the above is postulated on the vapor admixture occurring in relatively still air, as in a structure. Any factor that tends to produce an admixture of the vapors with air will alter the picture. A draft through a building is sufficient, or the activity of a machine may well serve to produce a wider explosive layer in the system postulated. A good illustration is the operation of the fan on an automobile engine when gasoline is leaking from a carburetor. Instances such as this which have been observed show that the fan on the motor mixes air fairly efficiently with the gasoline vapors, thereby increasing greatly the thickness of the explosive layer. The force of such explosions is much higher than ordinarily observed from gasoline vapor explosions, and the fire that follows may be correspondingly less, although this depends largely on the amount of leakage that has occurred.

The heaviness of organic vapors as compared with air must also be considered in connection with special circumstances of the environment. For example, machines are sometimes cleaned with hydrocarbon solvents in order to remove grease and other debris. Here, not only is there a high concentration of vapors at low level in the enclosure, but in some instances there are still lower regions to which the vapors may sink, such as a pit of some sort. If the vapors sink into this kind of a depression, it is in the pit that the hazardous condition is established, and any spark or flame in this region will cause an explosion, even though the region is physically separated by a substantial distance from the area in which the fumes were released. Drains, sewers, and similar low points are also favored from the standpoint of explosions somewhat remote from the area of release of vapors.

It can be stated that as a general rule the "explosion" of solvent or liquid vapors tends to produce rolling fires rather than violent explosions. The latter are more characteristic of substances that are lighter, such as natural gas. So many additional factors of local occurrence can modify this generality that the investigator must be alert in his evaluation. To assume that because one type of situation tends to be true, it will always follow is a mistake that can be made very readily, and will sometimes cause failure to diagnose properly the true cause and sequence of the effect.

One of the most significant factors in determining the character of the explosion is the degree of admixture with air, or where in the explosive range the percentage of vapor falls. This is the same issue discussed earlier in connection with "rich," theoretical, and "lean" mixtures. The reader is referred to Table 3 of Chapter 4 for details of explosive (or combustible) limits. In applying this consideration to an actual explosion, or a rolling fire, there are other details to consider. In an enclosed space such as a room or basement, where flammable vapors are present, the distribution will be far from uniform, as indicated earlier. Not only will the vertical distribution vary depending on the vapor density of the vapors or gases, but the longitudinal concentration may also vary, depending on time allowed for spread of the vapor and admixture with air and with presence of air currents, generally imperceptible but still effective.

Air currents are not only created by localized conditions of ventilation and by movement of machinery, persons, etc., but they are generated even by pilot lights or other larger flames. A pilot light creates small volumes of hot gas, which rise because of the reduced vapor density of hot gases; cool air flows into the base of the flame, giving rise to a slow but definite flow of air toward the flame. If heavy vapors are being released, even at some distance, this slow air flow will move the vapors toward the pilot and result in ignition. This is the main cause for ignition in explosions of vapors at a considerable distance from the main region in which the vapors exist. By contrast, an electrical spark, having only a short duration, will not give rise to the same effect. Either the vapors are in the explosive range in the neighborhood of the spark and will explode, or if outside that range, nothing will happen as a result of the sparking.

IGNITION

All of the considerations discussed with respect to ignition sources for fire are to some extent applicable to diffuse explosions. Since the phenomena are basically the same, the only serious difference is in the fact that the fuel mixture must be present at the source of the hot spot that ignites it in order to create the explosion. With a fire, continued heating of pyrolyzable fuel will create the conditions for the fire but not for the explosion.

Any hot spot or region, regardless of its size, can initiate an explosion of vapors if it fulfills the following criteria:

1. At some point, the heat is above the ignition temperature.
2. At this point, the gas mixture is within the explosive range.

These criteria include all flames, electrical sparks or arcs, including those created by static electric discharge, and hot solids such as heated metal. Interestingly enough, smouldering combustion, as on the tip of a lighted cigarette, is generally incapable of initiating the explosion. This effect is not simple to explain when a tiny flame will immediately initiate the explosion in the vapor mixture, but the cigarette will not, even when puffed and exposed to fumes continuously. It is apparently related to the ash covering that forms immediately over the glowing end which provides the same type of protection that is given by the wire gauze that covers the flame of the Davey safety lamp used by miners.

CONCENTRATED EXPLOSIONS

Explosions of materials in the category of concentrated or "high" explosives differ greatly from the diffuse explosion in various significant ways. Some of these have been indicated, particularly the volume effect. There are other differences that involve the thermodynamics of the system, and these will only be briefly discussed here. Whereas diffuse explosions are normally simple combustion processes carried out in a very short interval of time, concentrated explosions may involve a totally different chemistry. For example, nitrogen is not combustible; it would serve very well to extinguish fires if its physical form were more suitable for use in fire extinguishers. If liquified, it should be at least as useful in this connection as the carbon dioxide extinguishers so widely used. However, it is costly to liquify nitrogen as compared with carbon dioxide, and it is not used for this reason. Nitrogen also constitutes some four-fifths of the air volume and, therefore, serves to retard the intensity of fire because the fire is related directly only to the additional approximate one-fifth that is oxygen. If the air were all oxygen, the fire intensity would be vastly greater, as has been demonstrated when astronauts were burned to death in an all-oxygen atmosphere.

Interestingly enough, nitrogen compounds are probably the most common types of compounds utilized in the formulation of high explosives. This is because most nitrogen compounds are basically unstable from the standpoint of their compositional chemical energy. Nitrogen is an element that resists chemical combination with other elements, in a sense preferring to remain in elementary form. This means that the compounds of nitrogen are basically unstable and tend to rearrange explosively to yield elemental nitrogen and a great deal of energy. Since oxygen is the chief element with which nitrogen is combined, it is relatively simple to include such combinations along with carbon compounds; in the explosion, nitrogen is released from its combination with great energy generation, and the oxygen also combines with the carbon compounds to produce more energy.

In addition to nitrated compounds, several other reasonably common compounds or mixtures are capable of undergoing reaction at explosive rates and with generation of very large amounts of heat. When a compound is explosive, it will invariably contain some specific type of chemical configuration that confers on it the tendency to undergo explosive reaction. These have been termed explosophores (1) and include the following chemical groupings:

1. $-NO_2$ and $-ONO_2$ in organic and inorganic substances.
2. $-N=N-$ and $-N=N=N-$ in inorganic and organic azides.
3. $-NX_2$ where X is a halogen.
4. $-N=C$ in fulminates.
5. $-OClO_2$ and $-OClO_3$ in inorganic and organic chlorates and perchlorates, respectively.
6. $-O-O-$ and $-O-O-O-$ in inorganic and organic peroxides and ozonides, respectively.
7. $-C\equiv C-$ in acetylene and metal acetylides.
8. $M-C$ metal bonded with carbon in some organometallic compounds.

In addition to compounds that contain groups such as those illustrated, there are a variety of mixtures that are subject to explosion. These are generally low energy explosions, and they are more often encountered as prankster bombs than as seriously dangerous materials. A possible exception is black powder, a mixture of carbon, sulfur, and nitrate. Probably the most common material used in bombs is ordinary dynamite, which basically is compounded with nitroglycerine soaked into a solid carrier material which is often combustible and may itself have explosive properties, depending upon the type and rating of the dynamite.

The number of possible explosive materials that have been made and tested is so large that for the present purpose it is not considered necessary to do more than outline them briefly. Interest in the extensive possibilities of compounding explosives is more important to those concerned with ordnance and other military uses, and with specialized industrial application, than with the investigator of fires and explosions. For example, specialized requirements for explosives of a definite character often lead to blending of several materials to produce the exact result desired. Such specially compounded materials are of infrequent concern to the investigator, however important they may be for the specific purpose at hand.

PROPERTIES OF TRUE EXPLOSIVES

All explosive substances produce large amounts of heat and most of them also produce large quantities of gas which expands because of the heat. All explosive materials are sensitive to heat in that raised temperatures will cause explosion at some characteristic limit. Most explosives are also subject to detonation, i.e., mechanical effects that produce explosion. Thus, the *sensitivity* of the explosive to both such effects is an important consideration, especially when extensive handling of the material is necessary. Some explosives effectively just burn but at very high rates, while others do not burn but truly explode without any combustion process being involved. On the basis of such properties, explosives may be classified as follows:

Propellants or low explosives. These are combustible materials that carry both fuel and the necessary oxygen for their combustion. They do not actually explode in the chemical sense. Examples are black powder and smokeless powder (nitrocellulose).

Primary explosives or initiators. These materials explode or detonate when heated or subjected to mechanical shock. *They do not burn*, and the explosion does not depend on being confined. Examples are mercury fulminate, lead azide, and lead salts of picric acid. Such materials are commonly used in caps and primers to initiate a larger explosion. They are quite sensitive materials.

High explosives. Compounds in this class are generally exploded by detonation of materials discussed in the paragraph above. When suitably spread out, they may merely burn, but when confined and detonated they produce very powerful explosions with great brisant (shock). Examples include dynamite, T.N.T., picric acid, nitrocellulose, and many other materials both single and in mixture that are useful for industrial blasting, military purposes, etc. Their force depends not only on the quantity of the explosive used but also on the vigor of its detonation and the degree of confinement. Military and ordinance uses have led to a very wide variety of explosive materials and mixtures in this category, the variations being of significance in such matters as handling hazards, reliability of function, improved performance, and greater power. To the fire investigator these are all secondary considerations.

SENSITIVITY TO HEAT

All explosives have temperature limits below which they do not explode. However, the mechanism for inducing the explosion can provide heat enough to initiate an explosion if properly arranged. Likewise, explosives that are exposed to sufficiently high temperature will explode as a result of the heating. Another important effect of heat is in altering the rate of explosions (2). Some typical temperature data are shown in Table 1.

As pointed out by these authors (2): "It is generally considered that an

TABLE 1. TYPICAL TEMPERATURE DATA(2)

Compound	Temperature, °F.	Explosion Time, Seconds
Tetryl (trinitrophenylmethylnitramine)	507	No explosion
	545	1.45
	680	0.325
Black powder	599	No explosion
	662	12
Nitrocellulose (13.4N)	338	No explosion
	507	1.30
	676	0.074
Picric acid	500	No explosion
	662	1.48
Mercury fulminate	338	No explosion
	460	1.03
Lead azide	597	No explosion
	624	1.14
	680	0.560
Nitroglycerine	401	No explosion
	606	1.15
	502	0.217

explosion is preceded by a relatively slow reaction which increases more or less rapidly to explosive violence." Thus, the temperature is of importance in several different contexts in the matter of explosions. This led to the recent technique of Plouff (3) of preventing explosion of planted bombs by immersing them in liquid nitrogen. Even with proper detonation, which should produce sufficient local heat to lead to an explosion, the resulting explosive reaction is so weak as to be negligible.

COMMON EXPLOSIVES

While special requirements for military uses have led to a wide variety of special explosives useful for specific needs, the interest of the fire investigator who occasionally encounters explosives as a part of the overall fire picture is far more limited. Likewise, the matter of ammunition is in some respects different, because of its specialized use which is of interest to sportsmen and others, independently of ordnance applications. Such individual uses of explosives, which are not of interest to the fire investigator, will be omitted from detailed discussion. Commonly encountered explosives are discussed briefly below.

Aromatic Nitro Compounds

Of the various aromatic hydrocarbons that are readily nitrated with strong nitric acid, only toluene has the property of forming readily a trinitro compound that has both the desirable explosive and physical properties. It can be loaded into shells by pouring of the molten material; it is readily detonated, insensitive to shock, powerful, and brisant. The slightly simpler compound, trinitrobenzene, is more powerful but less convenient. Also, the next compound of the series, xylene, yields a trinitro compound that has a reduced force of explosion and is of less value. TNB, TNT, and TNX are all well-known explosives.

Dynamites

Dynamite is unquestionably the most available of high explosives. It is the one most often used for illegal bombing and also likely to be involved in fire investigation. There are several types of dynamite with different properties which should be understood.

"Straight" dynamite. This is essentially nitroglycerine absorbed in "dope" compound of combustible absorbent (wood pulp) and an extra oxidizing material (sodium nitrate) to which is added a small amount of an antacid such as calcium carbonate or zinc oxide.

Ammonia dynamite. Here, a considerable part of the nitroglycerine is replaced by ammonium nitrate. Often, it is protected from moisture by a coating of vaseline or paraffin and is usually neutralized by zinc oxide. It commonly contains less wood pulp than does straight dynamite and generally includes sulfur and cereal products.

Gelatin dynamite. This is a broad term including a variety of formulations in which the common factor is the substitution of nitrocellulose (smokeless powder) for the relatively inactive wood pulp. The mixture is a gelatinous, plastic mass, sometimes termed "blasting gelatin." It has considerably more force than ordinary dynamite because it is a mixture of two explosives.

Low-freezing dynamite. Dynamites in this category are modified for the purpose of reducing the freezing point of nitroglycerine. A number of compounds that are explosive will have this effect and may be substituted for some of the nitroglycerine. Included are such compounds as nitrotoluenes, nitroxylenes, nitrohydrins, nitrosugar, and nitropolyglycerine.

Dynamites all explode on overheating and exposure to heavy shocks as in detonation or being struck by a bullet, for example. They are far more stable to shock than is nitroglycerine, which is the major actual explosive substance in them. Dynamites are readily destroyed by spreading them out in the open in a thin layer and burning them. If properly done, the fire, though intense, is quiet and will not lead to explosion.

Other High Explosives

Additional common nitrated aromatic compounds include picric acid (2,4,6-trinitrophenol) and ammonium picrate, also known as "Explosive D." A variety of nitrated aliphatic compounds are also of great importance. Nitrocellulose (smokeless gunpowder) and nitrostarch, used as a blasting explosive, are perhaps most common. Tetryl (trinitrophenylmethylnitramin) is used primarily as a booster charge to transmit and intensify a detonating wave to the less sensitive bursting charge in high explosive shells. Mercury fulminate, generally mixed with potassium chlorate, is used in blasting caps because of its very great sensitivity to mechanical impacts and electric detonators. Many other materials are used in primers. Numerous additional explosive materials and combinations are in use that will not be listed here but can be found in standard references.

Detonators and Safety Fuses

Detonators require two types of explosive, one that is sensitive to the fire of a safety fuse which, in turn, will explode a second more powerful charge that detonates the main explosive. Lead azide is frequently employed for the first because of its high sensitivity, and a typical high explosive such as TNT or PETN is used for the detonating charge. Such detonators are designed to attach to safety fuses by placing the explosive materials inside a tubular copper or aluminum case into which the fuse if inserted and held by crimping of the end of the detonator case around it.

Many types of safety fuses are manufactured and used in different countries, the differences residing in diameter, number of wrapped layers, color and type of insulation of the outer layer, number of threads in the layers, identifying thread numbers and colors, and rate of burning. A core of black powder is wrapped successively with a standardized set of yarn or jute layers, ordinarily three in number. Tape is also used instead of thread, and the outside layer is insulated with tar, gutta percha, rubber, or plastic. Because each manufacturer uses a distinctive combination for each of his products, a piece of safety fuse may be readily traced if found intact. Age, condition of storage, and defects can all alter the rate of burning to some degree; pressure applied to the fuse while burning will alter its rate of burning in the compressed region. The finding of any fuse or detonators, intact or in fragments, will be of great value to the investigator of fires in which explosives may have been involved.

Other devices that are rarely encountered in fire investigation, but which may at times be of importance, are *detonating fuses* and *detonators for electrical ignition.* Both items are widely used by the military and in large blasting operations. Where a great deal of rock must be moved, for example, the use of simple detonators and safety fuses would be exceedingly slow. Industrial blasting is almost entirely done by means of electrical detonation at present, making possible any number of simultaneous explosions and maintaining safe distances as well.

Detonating fuses are made in a manner generally similar to safety fuses, and they are coded in a similar manner. Filled with the high explosive PETN instead of black powder, they burn at a rate of fifteen to twenty thousand feet per second. They are also very insensitive to blows or handling, thus imparting great safety to their use. They can be used to connect multiple explosive charges so that all will explode at essentially the same time. They are unlikely to figure in fire investigation.

Detonators for electrical ignition are in most respects similar to the simple detonators described, except that the heat that initiates explosion of a sensitive material in them is obtained from a small wire that is heated electrically while embedded in the charge. The possibility of finding such mechanisms in connection with fire investigation is somewhat greater than of finding detonating fuses, but it is not very probable.

DETECTION OF EXPLOSION IN A FIRE SCENE

If high explosives are used in connection with a fire, or if stored explosives are detonated as the result of a fire, it is probable that some witness will have noted the very sharp report of the explosion as contrasted with the normal noises of the fire, or even of diffuse explosion, which is duller in sound. If such observations have been made, it is sufficient reason to search for localized areas of extra damage not attributable to the fire itself.

Some fires are inherently very noisy and may be accompanied by numerous "explosions," e.g. fires in chemical plants where there are numerous solvent containers. Also, some fires are both large and isolated so that the presence of a special report due to high explosives will not be especially noted. In all such instances, the careful investigator is still likely to note effects of such an explosion, and he should be alert to them. They are different qualitatively from the effects of mere burning, in that the high concentration of energy at the point of the explosion will produce violent local damage that cannot be attributed either to the fire alone or to simple collapse as a result of weakening of a structure. In one large warehouse fire, on an upper floor there was a hole blown through the floor that did not fit a normal fire pattern at all. Not only were the floor planks broken sharply in a rectangular pattern, but a heavy beam supporting the floor was also sheared off quite sharply. The ends were jagged but abrupt, showing the effect of a very large and concentrated force that simply sheared out this section of floor. It could not be determined whether the explosion was a portion of an act of arson or whether explosives had been stored in this region, but only a localized high explosive in this region could have produced the observed effect.

INVESTIGATION OF EXPLOSION

When any localized region in a fire scene shows the results of extreme force, search for indications of the explosive must be undertaken. Perhaps this was an area in which explosives were stored and a reasonable conclusion is possible. Some explosives leave residues which can be seen on the surfaces around the region of the explosion. This is especially true of black powder and several chemical systems that include inorganic materials. Most high explosives are totally organic, and they leave little or no such residues. Search should be made for remains of any fuse or detonators that sometimes are not totally destroyed in the explosion. Finding such objects generally indicates that the explosive system was arranged and, therefore, a criminal act under the circumstances. If any unusual residue is located, it should be subjected to careful chemical and physical study in order to obtain from it all of the information possible, some or all of which will be useful in tracing its origin. This is not ordinarily the task of the fire investigator but of the explosives expert. Much can be done in such matters as identifying the tool with which a detonator was crimped, for example. Sales of dynamite can be traced often to the buyer, and such items as safety fuses are generally a very valuable source of information as to the origin of the material.

References

(1) Urbanski, T. *Chemistry and Technology of Explosives* (Vol. I). Pergamon Press, Macmillan, New York, 1964.
(2) Henkin, H. and McGill, R. "Rates of Explosive Decomposition of Explosives." *Ind. Eng. Chem.*, **44**, 1391, 1952.
(3) Plouff, L. Private communication.

17

Building Construction Materials

Although separate consideration has been given to various types of materials that are utilized in building construction, some have not been discussed and no overall consideration has been given to structural arrangement of materials. Non-flammable as well as flammable items have a direct influence on the development of a fire, and these have received no special attention in this volume. It is convenient to divide structures into portions for separate consideration, because the flammability and fire hazard relate in one instance to one portion of the structure and in another instance to some other portion.

STRUCTURAL SHELLS

This term refers to the outside or supporting portion of a structure, whether it is residential, business, or commercial. It includes especially the outside walls and supports and occasionally wells or similar structures at the center of a building which are increasingly being used in modern architecture as supporting structures. Such shells can be considered to fall into the following categories:

Masonry. In this category are included concrete cast and block, tile, brick, and similar materials, possibly even adobe. Shells constructed of masonry are clearly not flammable and will not directly contribute to a fire. To assume that they are not subject to fire loss is a mistake, because within such shells there are ordinarily walls, floors, and other structures of wood, flammable furnishings, and other flammable materials. A fire that develops within such a masonry shell may actually be more intense than if the walls are directly involved, because of the insulation effect and chimney actions which may be present. Some of the worst conflagrations have involved this type of structure.

Wood. In some portions of the world, wooden shells for residences are the rule, and in nearly all parts some residential buildings at least are basically of wood. Business and industrial buildings are less commonly constructed in this manner at the present time, but many older buildings still in existence were so constructed. These may be referred to as "fire traps" and often they are, but the reasons relate more to the interior construction and content than to the shell, which is rarely the point of fire initiation, regardless of what material is used for its construction.

A special case under wood shells is that in which the exterior is some form of plaster such as *stucco*. The presence of such material on the outside of the exterior walls has little influence on the fire hazard, except that which exists because an adjacent building is burning. In such an instance, there is less likelihood of the stucco-coated building becoming ignited. However, interior fires, which make up the vast majority of all fires, will be quite independent of the presence or absence of a stucco coating.

Glass and metal. Some modern buildings have shells that are virtually all metal, generally steel or aluminum, and glass. These, like masonry buildings, have no intrinsic fire hazard from the shells, but interiors and contents can burn readily if composed of basically flammable material.

INTERIOR STRUCTURE

The materials and arrangement of flammable structural materials on the inside of a building are far more important as related to their fire hazard than is the shell. If walls, floors, doors, window frames, and the like are predominantly of wood, the fire hazard in such a building can be very great but is altered by various other considerations, some of which follow:

Floors. In multistory buildings especially, the material of which the floors are constructed is often critical. Wooden floors, however they may be finished—polished in the residence and thick rough planks in the warehouse—are likely to be breached in any major fire, spreading the flames upward from floor to floor and limiting the possibility of isolating the fire before it grows too large. Concrete floors, frequently used between stories in industrial and some other buildings, tend to prevent spread of fire from story to story and to make the containment of the fire much easier. The bottom floor of a building is of less consequence than the others, because in most instances it is not breached by the fire and serves only as the low point for the collection of rubble. Exceptions exist to this rule, however. Some residences built on steep ground may have an elevated lower floor; even on flat land, the first floor is sometimes raised sufficiently to allow fire to develop below it. Such a situation is especially troublesome in cases of arson, but other causes for such a fire are encountered. Gas may escape under the floor, and trash or grass fires that are too close to the building may also get under it. The common wood floor that is raised from the ground can thus constitute a fire hazard, even when it is the bottom floor.

Ceilings. In multistory buildings, the same considerations for floors apply also to ceilings because the ceiling of one floor is the lower side of the floor in the story above. One very important difference exists, however. Because fires burn upward preferentially, the ceiling is always exposed to fire when any substantial fire exists within the room. If the ceiling is non-combustible, as with sheetrock, the fire is inhibited from burning through to the story above. The situation with floors is different, in that the fire burns away from the floor and will rarely breach it, except when falling, burning material comes into contact with it. Thus, it is far more important that the ceiling be fireproof than that the floor be non-combustible. When both can burn, it is evident that the situation is one that encourages spread of flame from story to story by burning through the combination.

Walls. The inner surface of exterior walls and both surfaces of interior walls are especially subject to attack by fire, both because of their vertical orientation and because of their large surface. Next to ceilings, they constitute the most vulnerable portion of the main structure of any building. With the exception of masonry walls, it will be normal for the main structural components, such as studding, to be of wood and therefore combustible. However, these structural elements are covered and often immune to attack if the covering itself is non-combustible.

Wall surfaces were traditionally of plaster, which was itself non-combustible, but was subject generally to cracking and flaking so as to expose layers of materials under the plaster. Often these were wooden lath which burn vigorously. Many older buildings remain which are finished in this manner and, even though the plaster will not burn, it also will not protect underlying fuel of the wall for more than a short time. Such walls were often papered, which again added a new component to the problem of fire resistance. A single layer of paper was certainly not especially hazardous; it would quickly char without adding much fuel to the flame and was, thereafter, relatively inactive in the fire. When the wall was repeatedly papered without removing old paper, the multilayer surface that resulted naturally created an increased hazard. Both the paper and adhesive with which it was attached were combustible, and thickening the layer increased the amount of possible augmentation of the fire. Paint used similarly had generally less effect, although thick layers of paint also have been shown to be considerable contributors to the intensity of a fire that involves them.

The use of various types of wall board as a substitute for plaster is a more recent development of wide application. Early wall boards were generally quite flammable and dangerous from the standpoint of fire, as well as being somewhat less than fully satisfactory for wall finishing. The introduction of gypsum board, or sheetrock, is possibly the most important development in developing fire resistance in low- and medium-cost housing that has so far occurred. Not only is it non-combustible, but it will resist fire for considerable periods, thus allowing containment of local fires. Every fire investigator will be struck by the fact that in a building in which extensive burning has occurred, those rooms that are lined with gypsum board invariably show the least destruction.

Wood surfacing, which was originally standard in nearly all primitive construction, was replaced by plaster and other materials but finally has again become popular because of the aesthetic appeal of the many manufactured plywood and veneer boards that are currently available. These materials are relatively inexpensive, highly decorative, readily applied, and durable. They can also raise the fire hazard by a very significant amount. If ignited, they tend to burn rapidly because of their thinness. They allow penetration of the fire into the interior of the wall, where spread can be on both sides of the board. And some plyboards have been known to delaminate, with separation of the very thin layers from which they are constructed. When this occurs, the greatly increased surface available to the flames causes very intense fires. It is not likely that materials so useful as these will pass rapidly from the building picture, but it can certainly be recommended that they be placed over a layer of sheetrock which at least prevents the burning on both sides and keeps the fire restrained, both in total intensity and from entering the interiors of walls.

Doors, windows, and trim. As a rule, doors and their frames, window frames, moldings, and other trim are made of wood and are therefore flammable. These items invariably augment a fire; the degree of their participation is proportional to the surface they expose to the fire. It is evident that the more general use of metal window frames will diminish the effect, but it is unlikely that doors and door frames will rapidly be replaced by non-combustible materials.

Combined with other wooden interior finish, these items will generally be a large determining factor in the fire intensity within a room that is afire. It is normal for some ten percent of the interior wall surface to be composed of such wooden material. When the percentage rises much above this, the fire hazard is raised proportionately.

Roofing materials. Most roofing materials in common use are relatively fire resistant. Tile and other ceramic roof is totally resistant; composition material, as well as tar and gravel, is very resistant; only shingles and shakes present any real hazard in connection with fire. Even these are not simple to kindle but, like all wood products, can catch fire when exposed to appropriate ignition agents. Contrary to some uninformed beliefs, asphaltic materials used in roofs are almost never a causal factor in the initiation of fires. They are so difficult to burn, in fact, that it requires generally a very intense building fire under them to burn them. They will often melt and flow, and in a heavy initial fire they will burn but only with reluctance and generally with a self-extinguishing flame. When wood can be avoided in the roof, there is little if any significant hazard of roof fires.

Supplemental interior structures. Every residence and most business and commercial buildings contain a rather large number of cabinets, cupboards, wardrobes, shelves, and the like. These are commonly constructed of wood and are by far more susceptible to fire and ignition than almost any portion of the main structure. Not only do they present a large surface which is combustible, but they contain and carry all types of flammable materials, such as clothing, paper products, volatile flammables, and other intrinsically hazardous materials. It is fair to say that a very large proportion of building fires are initiated in these supplemental interior structures. In view of their extraordinary susceptibility to fire risk, it is curious that little attention seems to have been given to making them more nearly fireproof. Wardrobes are the one possible exception, since many of them are lined with sheetrock. Cupboards, cabinets, and shelving are almost always exposed and intrinsically dangerous from the fire standpoint. More general use even of such materials as micarta, which is far less subject to ignition than is wood, would diminish fire risk appreciably.

Non-structural furnishing. Although the builder will rarely have any control of such items as furniture and drapes, this category of items cannot be ignored in any discussion of fire risk in environments used as living quarters. The initial ignition of more fires probably occurs in these items than in any other single category. Drapes are especially susceptible to easy ignition, except when they are constructed of relatively non-flammable material such as fiber glass, some synthetics, or wool. Any carelessness in lighting a cigarette close to a drape can readily cause the loss of a house. Wind can blow them into a cooking fire or place them on an ash tray containing burning cigarettes. Furniture is less easily ignited by carelessness than are drapes, but a very large number of fires start in a bed or on a chair or davenport. If furniture or drapery become aflame, the effect of this fire on adjacent permanent structures that are flammable is obvious. When the surrounding permanent structure is relatively fireproof, the fire can be readily localized and extinguished before more than minor damage is done.

STRUCTURAL ARRANGEMENTS

Virtually every structure will contain within it enough combustibles to allow a fire to occur. The arrangement of them will have generally far more influence on the speed and course of the fire than will the materials themselves. Such architectural considerations may be divided into two directions of interest: (1) retardation and containment of the fire and (2) protection of occupants. To a degree, the two points of view are interdependent because any measure that contributes to one will inevitably have an influence on the other.

Retardation and containment of fire. This end will be accomplished to a degree by attention to the following items:

1. Exposure of a minimum of combustible material such as wood at surfaces.
2. Avoidance of open chimney-like structures, such as open stairs, dumb waiters, and ventilator shafts of any type, that are not protected by ceramic or metal interiors and by doors, preferably fire resistant.
3. Interior walls that burn through rapidly because of the material of their construction.
4. Careless or inadequate sealing of such items as lighting fixtures that cause breaches in walls or ceilings for installation of the fixture and passage of wiring.
5. Presence of transoms between rooms and hallways. These are always high, in the region of fire spread along hallways, and open the rooms to this fire.
6. Presence of sprinkler systems, generally limited to industrial and business buildings for aesthetic and economic reasons.
7. Presence of alarm systems.
8. Availability of fire extinguishers at strategic locations.
9. Proper installation and shielding of electrical and gas appliances, wiring and piping.

Protection of occupants. In considering this important question, it must be remembered that occupants are more often overcome by carbon monoxide than by fire itself. Their protection, therefore, must be primarily built around this hazard rather than the hazard of being burned directly. Clearly, any measure that retards or contains a fire will have an influence on the protection of the occupant, since the fire will be less likely to become general and widespread. In addition to the items listed under the above category, the following must be included:

1. Presence of multiple escape routes, at least one of which is likely to be away from rather than through a possible burning area. Fire escapes, windows large enough for easy exit, and similar arrangements fall in this category.
2. Tight doors between bedrooms and hallways. Open doors or doors that have significant gaps around them provide easy entrance for lethal quantities of carbon monoxide to the room in which someone may be sleeping. This is perhaps the most important of all internal construction details that will protect the occupant from the most important hazard of the fire to his life, carbon monoxide.
3. Easy means of communication, such as telephones, in bedrooms. This may serve as a twoway protection, allowing an outsider to awaken the sleeping occupant or allowing the occupant who becomes aware of a fire to communicate with the outside without exposure to the fire itself.
4. Adequate means of exterior ventilation of sleeping rooms. Ventilation within the room will constantly diminish the carbon monoxide content and increase the time available for escape. Interior ventilation will, of course, have the opposite effect, allowing gases created by the fire to circulate freely into the sleeping quarters.

It is clear that no completely acceptable construction materials or arrangement is likely to be available at the present time for the prevention of damaging fires. It is entirely feasible to build a structure that is essentially fireproof and totally safe by utilizing such materials as masonry, metal, fiber glass, glass, and fire-resistant plastic. Even here, there will inevitably be items of clothing, paper, and other combustibles, that are so much a part of modern living that fires can still occur, although in a more localized and less dangerous context. When other considerations of aesthetics and comfort are considered, it is highly unlikely that humans will soon make the necessary effort to immunize themselves to the effects of fires. At the same time, both builders of structures and the persons who occupy the structures could do a great deal to minimize the dangers of fire without greatly detracting from serviceability, attractiveness, and comfort. The worst situation is the one that develops when the builder deliberately introduces extra hazard as a means of reducing costs. Probably the only answer to this unfortunate possibility lies in strengthening the building codes and especially in informing the public sufficiently that they will take an active interest in the method of constructing the houses they buy and occupy and its relation to the potential fire hazard. Naturally, since most fires result from carelessness and ill-considered acts of individuals, it is especially important that the public be made aware of the serious consequences of any carelessness with the great destroyer, fire. In the meantime, it is unlikely that the need for efficient and high-grade fire investigation will diminish appreciably within the foreseeable future.

Supplemental References

Comparative Burning Tests of Common Plastics. Underwriter's Laboratories, Inc., Bull. of Research No. 22, Aug. 1941.
Fire Exposure Tests of Ordinary Wood Doors. Underwriter's Laboratories, Inc., Bull. of Research No. 6, Dec. 1938.
Fire Hazard Classification of Building Materials. Underwriter's Laboratories, Inc., Bull. of Research No. 32, Sept. 1944.

1 appendix

Fire Experimentation

This appendix is devoted to indicating the potentiality of direct experiment with a minimum of specialized or elaborate equipment and under the most practical conditions available to nearly all investigators. Professional and commercial laboratories devoted to study of ignition and combustion of materials must do comparative testing under highly standardized conditions; their results often do not answer the simple questions that must be resolved in many practical situations.

The small-scale experiment may not appear relevant to the large fire, but it must be considered that nearly every large fire started as a very tiny fire, and the location and conditions of this tiny fire were the prime concerns of the investigator. In fact, fire origin can rarely be tested on a large scale because fire origins are not normally large. Thus, this type of testing is actually most relevant to the basic purpose of the fire investigator.

Not all of the matters of interest are illustrated in this appendix, nor does it more than indicate the breadth of experimentation that has been necessary over the years of fire investigation. Rather, it illustrates how simple experiments can be made to yield the necessary information for application to practical problems. Many more experimental findings are indicated from point to point in the text. The ingenious investigator can readily test these and other situations under conditions that are controllable and inexpensive. The results of such simple experimentation will very often be critical in the final interpretation of the origin and cause of the fire, whatever its size or destructiveness.

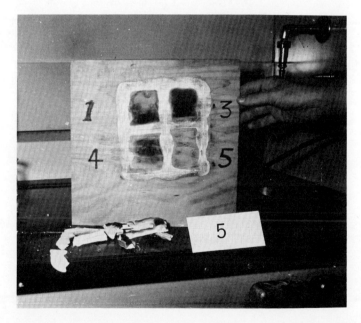

Figure 1. *Scorching of soft woods due to liquids burned on plaster dams. (1) Jasco brush cleaner. (3) Radiant paint and varnish remover. (4) Hercules turpentine. (5) Energine. It will be noted that different liquids have a considerable difference in the amount of charring they will produce in wood presumably due to the greater penetration of the wood by some than by others.*

Figure 2. *Cleaning solvent hazards.* (a) Energine burning on hardwood flooring, left. (b) Left showing the discoloration produced by burning Energine on the floor as compared with the deep char produced by small blazing curtain dropped on the right side of the same flooring.

Appendix 217

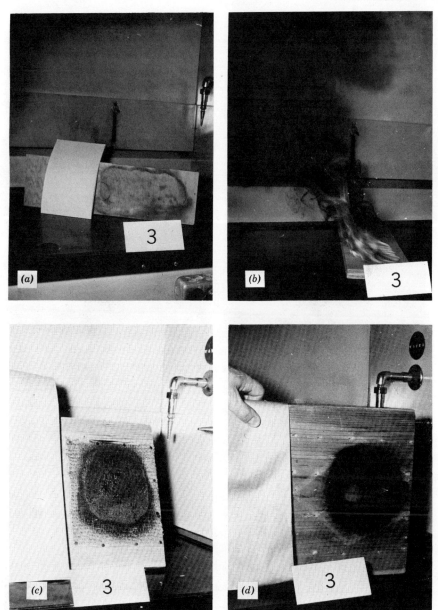

Figure 3. *Hazards of paint removers. (a) Radiant paint remover burning on soft wood board. (b) Damage to board after burning of paint remover. (c) Radiant paint and varnish remover effect when burned on carpet. (d) Effect on wood below burned carpet after removal of latter. It will be noted that some paint removers are quite flammable and productive of vigorous flaming.*

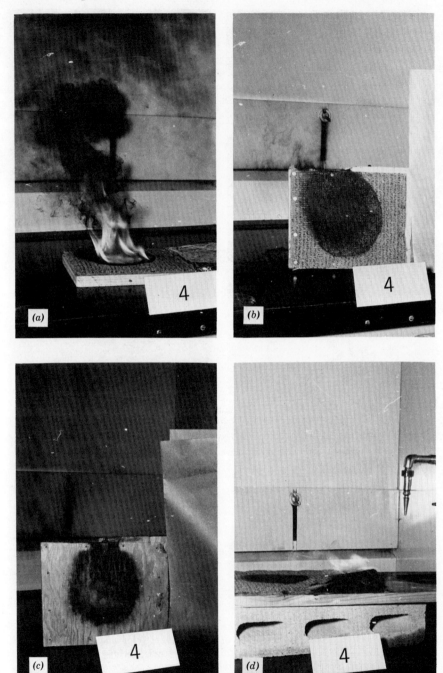

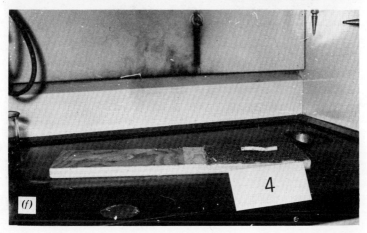

Figure 4. *Turpentine as a fire hazard. (a) Turpentine burning on carpet. Note very black smoke produced. (b) Effect of burning turpentine on the carpet. (c) Board under carpet after removing latter. Note the charring effect due to greater penetration of turpentine into the wood than with other solvents. (d) Turned back edge of carpet soaked with turpentine 36 hours prior to ignition. Note the persistence of turpentine in fabric materials and wood. (e) Charring at turned-back edge of carpet. Note the unusual effect of flammable solvents in the case of turpentine due in this instance to flames playing on the wood as well as the soaking of the wood with the material. (f) Cigarette burned on the carpet moistened with Hercules turpentine. Note that no fire is produced by a cigarette placed directly in contact with a material as flammable and otherwise hazardous as turpentine.*

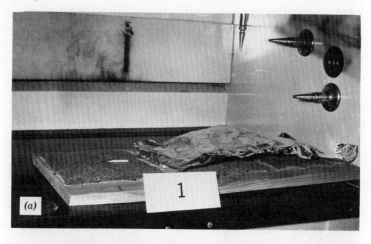

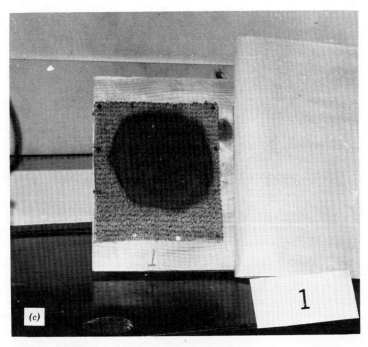

Figure 5. *Sequence—fire hazard from brush cleaner. (a) Jasco brush cleaner would not ignite after 1/2 hour standing on the carpet. (b) Freshly applied Jasco brush cleaner burning. (c) Results of burning Jasco brush cleaner on the carpet.*

222 Fire Investigation

Figure 6. *(Left)* Hole burned completely through soft wood by a blazing curtain. *(Right)* Superficial surface scorch produced by burning Jasco brush cleaner on the board.

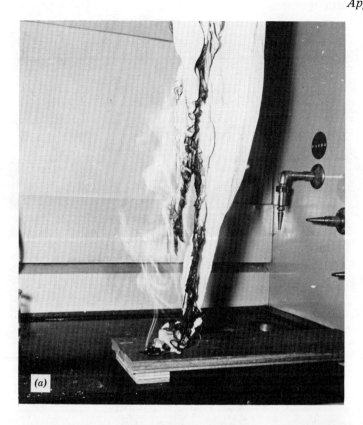

Figure 7. *(a) Small curtain burning over hardwood flooring. (b) Char produced by blazing curtain droppings. Note the deep charring effect even on hardwood of blazing material.*

224 Fire Investigation

Figure 8. *(Center) Deep charring due to a hot coal on wood. (Left) Hole burned in carpet by hot plaster. (Right) Deep hole burned in carpet by hot coal.*

Figure 9. *Hot plaster burning into carpet. Plaster has been heated over a charcoal fire.*

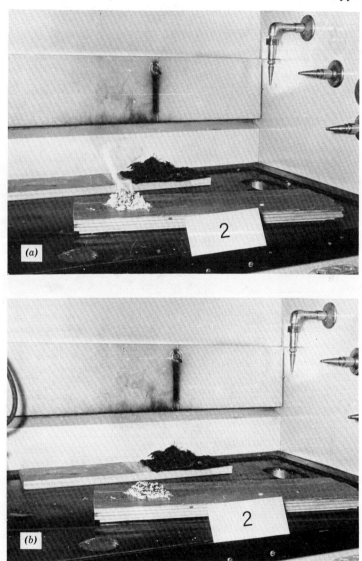

Figure 10. *Flammability of paint removed with paint remover. (a) Paint removed with Jasco paint remover burning. (b) Remains after burning. Note that the paint did not burn with any intensity, and Jasco paint remover was itself not flammable after standing for a short time.*

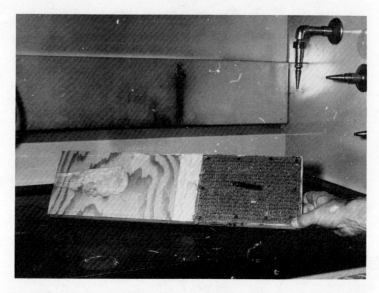

Figure 11. *Effect of complete burning of a cigarette lying on carpet. Note that only a charred pattern of cigarette remains.*

Figure 12. *Cavity burned (melted) and charred (finished) in sponge rubber by total combustion of cigarette lying on it.*

Figure 13. (a) Cigarettes burning on fabric placed on top of sponge rubber. (b) Effects on the fabric and the sponge rubber.

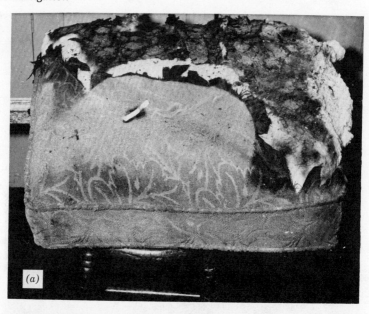

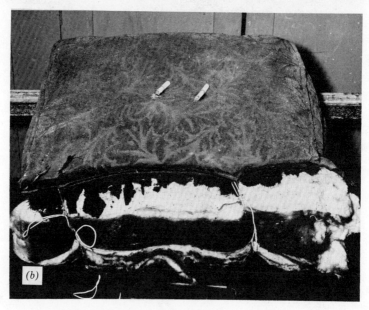

Figure 14. *Effects of burning cigarettes on upholstered cushions. (a) Surface charring only. (b) Note that flaming fire will not result, but that smouldering fire may penetrate into the underlying padding.*

 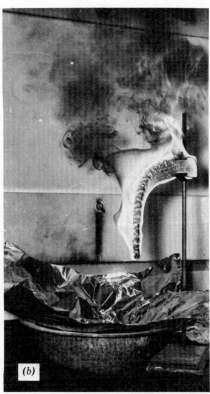

Figure 15. *Sponge rubber as a fire hazard. (a) Sponge rubber in an early period of combustion. (b) At a later stage of same combustion. Note the very dense black smoke produced, and the fact that the material burns freely.*

230 Fire Investigation

Figure 16. *Experiments with carburetor fire. Carburetor filled with gasoline would not support a fire until it was placed in a pool of gasoline and burned as shown. (a) Start of the fire. (b) Fire at a later stage. (c) Undamaged carburetor after the fire had extinguished itself.*

Appendix 233

Figure 17. Sequence showing effect of cigarette in creating a smouldering fire in saw dust. (a) Lighted cigarette on saw dust. (b) Beginning of smouldering fire. (c) Expansion of smouldering area. (d) Smouldering has occurred throughout the entire pile with a sharp demarcation between the charred and uncharred saw dust.

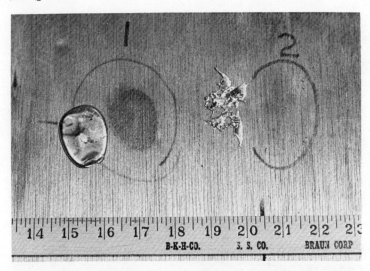

Figure 18. Effects of Plumber's 50-50 solder in starting fire. (1) Effect of pouring melted solder on thin plywood. (2) Effect of dropping melted material from a few feet height on same board. (3) Melted solder dropped for a few feet on paper packing and excelsior. In no instance was there more than a trace of scorching, even at melting temperature.

2 appendix

Illustrations of Fire Origins

However many thousands of photographs of fire scenes are taken, it is rare for a single one of them alone to illustrate adequately the effects that are observed and correlated with each other in the final interpretation of a fire origin or cause. The ultimate reasoning usually results from consideration of many factors in the fire, which require many photographs to illustrate.

Another difficulty of the photograph derives from the fact that no detailed information is obtainable from a single photograph that reveals the entire fire scene. Thus each photograph shows only a small portion of the total number of factors that are important. With this limitation in mind, a few photographs that show some clear cut information relative to fire origin are included in this appendix. They will not substitute for actual experience and the many hours that may be expended in studying the remains of a single fire before the entire sequence and cause become clear and unassailable.

Figure 1. *(a) A very heavy supporting beam burned totally through by blow-torch effect from a spray paint thinner in burst drum. (b) The appearance of this drum is entirely typical of a sealed vessel containing liquid which has been heated in a fire.*

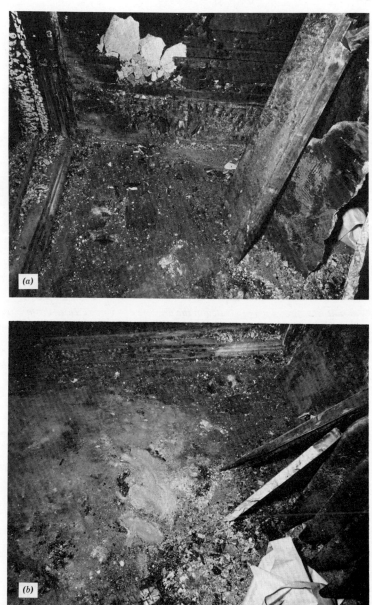

Figure 2. *(a) and (b) Localized burn including holes burned in floor from flaming draperies.*

Figure 3. *Result of a fire which originated in the drawers of a bureau. Note the localization of fire in the immediate neighborhood of the point at which the bureau stood, both on walls and floors.*

Figure 4. *An arson committed by pouring solvents under the location of a gas heating appliance. It will be noted that the floor at the upper right is not burned, and that the burning of timbers below the floor is extensive.*

Figure 5. *Bottom shelf of a cupboard in a house in which a serious fire had occurred. Some liquid flammable had spilled in this cupboard which shows burning to floor level and the outlines of containers on the top.*

Figure 6. *Open gas line under a roof structure. Note the fact that the gas flame has locally burned away the structure above it.*

Figure 7. *Results of a gas explosion in an attempted suicide. Although the force of the gas explosion is greatest at the top, the difference in strength of the walls top and bottom will frequently produce a displacement of the wall at the bottom.*

Figure 8. *Three photographs showing highly localized and very low burns produced by accelerants burning on a grout floor. Note severe attack on timbers down to the bottom of the wooden portion of the wall.*

244   *Fire Investigation*

Appendix 245

Figure 9. *Three photographs showing different points of origin of a fire in rooms considerably separated from each other in a business building. Flammable solvents had been used with burning of timbers to floor level in all instances.*

Figure 10. *Three photographs showing a weed burner and its effect when used to initiate a fire on logs supporting sawmill machinery. Number three of the sequence shows the deposition of soot on the steel parts immediately above the point of burning. This soot resulted from oil combustion.*

Figure 11. *Sequence of two photographs illustrating effect of lightning on a power line crossbar which did not initiate a fire. Note the explosive forces generated in the wood due to rapid release of gases on heating by the lightning bolt.*

Figure 12. *Effect of motor fire on adjacent aluminum wheel in electric clothes dryer.*

Figure 13. *Effect of water spray on hot concrete to produce spalling. Origin of this fire was close to the region in which the wall showed the effect. It was a warehouse in which a great deal of material and a number of vehicles had been stored.*

Index

Accelerants, identification of, 152
Acetate fabrics, fire hazard of, 127
Alcohols, accelerants in arson, 168
Aluminum, as fuel, 58
Ammunition, fragments as fire cause, 110
Animals, role in starting fires, 113
Appliance condition, investigation of, 142
Appliance cords, as fire cause, 103
Appliance wiring, 104
Appliances, as fire cause, 93
Aromatic compounds, 8, 44
Aromatic nitro compounds, as explosives, 202
Arson, access to premises in, 160
  chemical ignition in, 171
  choice of site for, 162
  concealment by appliances, 162
  corpus delicti, 176
  defined, 158
  detection of, 138
  fire ignition, 169
    by burning glass, 172
    by glowing wire, 171
    by sparks, 170
  history of legal definition, 174
  intent in, 158
  legal aspects of, 173

Arson, liquid flammables in, 164
   magnitude of fire in, 161
   model law of, 175
   modus operandi of, 160
   multiple fire origins, 165, 177
   necessary factors of, 161
   smoldering materials in, 169
   trash accumulations in, 163
   types of fuel in, 167
   use of cigarettes, 170
   weed burner in, 169
Arsonists, types of, 159
Asbestos fabrics, fire hazard of, 128
Asphyxiation, carbon monoxide, 181
Attic fires, 77
Automobile exhaust, source of carbon monoxide, 188
Automobiles, fires in, 119
   electrical origin of, 122

Benzene, 8
Boats, fires in, 123
Boiling point, 37
   of materials, 38
Bonfires as fire cause, 109
Breaking and entering, examination of, 156
British Thermal Unit, definition, 42
Building materials, cellulosic, 52
Burn, legal definition of, 177
Burned materials, examination of, 154
Butane, 7

Carbohydrates, 9
Carbon, oxidation of, 6
Carbon dioxide, formation of, 6
Carbon disulfide, explosive limits of, 35
Carbon monoxide, asphyxiation, 181
   investigation of, 190
   nature of, 182
   blood saturation with, 184
   dangerous concentration of, 182
   formation of, 6
   indications of generation of, 188
   rate of absorption, 184
   sources of, 185
   types of hazard, 183
Carboxyhemoglobin, definition of, 182
   stability of, 182
Carburetors, automobile, as fire hazard, 120
Ceilings, construction of, 208

Cellulose, 10
   nitrate fabrics, fire hazard of, 128
Char, origin of, 13
Charcoal, 49
Charring, depth of, 80
Chemical reactions, in fire ignition, 93
Chimneys, in fire causation, 107
Cigarettes, fire ignitors in arson, 170
   in starting fires, 111
Cloth, flammability classification, 130
Cloth flammability, laws regulating, 130
Clothing fires, 125
   effect of cloth type, 126
   effect of design, 129
Clothing, ignition of, 129
Coal, composition of, 47
Coatings, effects on fire pattern, 77
Coke, 50
Combustion, chemistry of, 4
   explosive, 17
   heat of, 42
   properties of paper, 52
Conduction, 19
Construction materials, building, 206
Convection, 19, 21
Cotton fabrics, fire hazard of, 126
Crown fire, 83
Crude oils, composition of, 43

Detonators, 203
   electrical, 204
Diffusion, of gases, 39
Direction of burning, factors that effect, 84
Dynamite, as explosive, 202
   types of, 202

Electric circuits, basic principles, 99
   light bulbs, fire hazard from, 117
   motors as fire cause, 104
   wiring, arrangement of, 100
      investigation of, 141
Electricity, loads and overloads, 103
Environmental conditions, effect on fire, 24
Exothermic reactions, 5
Experimentation, in fire investigation, 156
   small scale, 214
Expert qualifications, in arson, 178
Explosion, 17
   association with fires, 193
   concentrated, 194

Explosion (Cont.)
    nature of, 198
    cryogenic prevention of, 201
    detection at fire scene, 204
    diffuse, 193
        ignition of, 197
        nature of, 194
    investigation of, 205
    vapor, 196
    types of air admixtures, 197
Explosive compounds, 199
    limits, 35
    of materials, 36
Explosives, common, 201
    high, 200
    nature of, 199
    primary, 200
    reaction time of, 194
    reaction volume of, 194
Exterior fires, interior fires from, 78

Fabric fires, 125
Fabrics, melting of, as burn hazard, 133
Fingerprints, significance in investigation, 156
Fire, kindling in arson, 168
    necessary conditions for, 15
    origins, 235
    patterns, deviation from normal, 73
        outdoor fires, 82
        roof and attic fires, 78
        structural fires, 71
        tracing of, 79
    retardation and containment in structures, 211
    retarding properties, testing in fabrics, 132
    spread, 72
        rate of, 80
Fireplaces, as fire cause, 107
Fires, sun-kindled, 86
Flame, color, 59
    effects of metals, 60
    gaseous reaction, 15
    temperature, 48
Flaming fire, 14
Flares, in automobile fires, 123
Flash point, definition of, 30
    of materials, 30
Floor burns, 74

Floors, burned holes in, 75, 76
    construction of, 208
Fragments, hot or burning, 105
Friction, as fire cause, 92
Fuel, air ratios, 11
    compounds, 7
    line leaks, as fire hazard, 121
    quantity of, in arson, 166
    mixture, lean and rich, 36
    state of, 12
    suitable for arson, 167
    tanks, automobile, as fire hazard, 120
    wetness effect, 26
Fuels, non-hydrocarbon liquid, 45
    types of, 29
Furnace backfire, 97
Furnishings, relation to fire hazard, 210

Gas, role of appliances in starting fire, 95
    leakage of, 94
    lines, as fire cause, 94
        failure from heat, 95
        mechanical fracture of, 94
    liquid petroleum, 98
Gaseous fuels, layering of, 40
Gasoline, 9, 43
    composition of, 43
    cracked, 44
Glass fiber fabrics, fire hazard of, 128
    in construction, 207
Glowing fire, 14
    nature of, 16
Ground fire, 83

Hardwood, 49
Heat, conduction of, 20
    nature of, 19
    relation to fire, 18
    sensitivity of explosives, 200
    transfer of, 19
Heating appliances, venting of, 187
    equipment, factors influencing operation, 187
    source of carbon monoxide, 186
High explosives, oxidant self-contained, 194
Hot objects, in fire ignition, 92
Humidity, effect on fire, 25
    factors influencing, 25
    nature of, 24

254 Fire Investigation

Hydrocarbon, 7
 detector, 137
 fuels, 42
Hydrogen, chemical behavior, 5

Ignition sources, 89
 temperature, definition, 32
  effect of pyrolysis on, 68
  of materials, 33
  wood, 51
Ignitors, primary, 90
Industrial processes, carbon monoxide from, 190
Interior structures, supplemental, 210
Investigation, determination of purposes of, 134
 equipment for, 137
 photography in delayed, 149
 relation to clearing of area, 136
 routine of, 138
 time element in, 135
Isobutane, 7

Kapok, fire hazard of, 126
Kerosene, 44
 in arson, 167

Laboratory examination, 152
Light bulbs, as fire cause, 113
Lighters, in fire kindling, 91
Lightning, as fire cause, 113
 effects of, 114
Linen fabrics, fire hazard of, 126
Liquid fuels, pyrolysis of, 65
Liquids, flammable, detection of, 146
 use in arson, 164
 vaporization of, 12
Low burns, significance of, 74

Magnesium, as fuel, 58
Masonry in construction, 207
Matches, in fire kindling, 90
Metal, in construction, 207
Metallic fiber fabrics, fire hazard of, 128
Metals, as fuel, 57
 as ignitors 109
 as indicators of fire temperature, 145
Methane, 7
 oxidation of, 11
Molotov cocktail, in arson, 166

Naphthenes, 44
Natural gas, 42

Occupants, protection of in fire, 211
Octane, combustion of, 11
Ohm's Law, 99
Olefins, 44
Organic structure, relation to pyrolysis, 67
Outdoor fires, causes of, 85
 investigation of, 84

Paint thinner, in arson, 168
Paints, combustion properties of, 57
 pyrolysis of, 69
Paper, and cloth, examination of burned, 155
 combustion properties of, 52
 ignition temperature of, 53
Paraffins, 44
Petroleum, 42
Photography, at fire scene, 147
 color, 148
 time of, in relation to fire, 147
Plastic fuel containers, in arson, 166
Plastics, behavior on heating, 54
 fire hazard of, 53
 ignition temperatures of, 56
Plywood, 51
Power lines, as fire cause, 86
Propellants, 200
Pyrolysis, 10, 47, 63
 nature of, 64
 without combustion, 69

Q10 Value, 18

Radiant heat, in fire transfer, 93
Radiation, 19
 effect in fire, 23
 nature of, 22
Rayon fabric, fire hazard of, 127
Redwood, burning properties of, 49
Roof fires, 77
Roofing materials, 210
Rubble, value in investigation, 144

Safety fuse, 203
Services, role in starting fire, 93
Ships, fires in, 123
Shortcircuits, production of, 102

Smoke color, 61
  relation to fuel, 62
  production of, 60
Smoking, as fire cause, 111
  in bed, as fire cause, 111
Smoldering fire, 14
Softwood, 49
Sparks, classes of, 105
  in fire kindling, 91
  travel of, effect of wind, 106
Spontaneous combustion, 19, 47, 114
  as fire cause, 86
  role of microorganisms, 116
  types and nature of, 115
Structural arrangements, 211
  for occupant protection, 212
  fires, carbon monoxide in, 190
  shells, 206
Structure, interior, 207
Sulfur, combustion of, 6
Synthetic fabrics, fire hazard of, 127

Temperature, effect on fire kindling, 23
  self ignition, 50
Terrain, effect on outdoor fires, 83
Textile fabrics, treated, 128
Textile fires, effect of spinning, 128
  effect of weaving, 129
  flammability, testing of, 129

Thermal conductivity, definition of, 20
  of materials, 21
Thermoplastics, burning rates of, 56
Toluene, 8
Trash burners, in fire causation, 109
  piles, flammable liquids in, 163

Vapor density, 39
  detector, 146
Venturi, adjustment of, 187, 188
  definition of, 186
Victims, bodies of, 151

Walls, construction of, 208
Waste baskets, characteristics of fire
  from, 143
Water, formation in fire, 5
Wind, effect on outdoor fires, 83
Witnesses, interrogation of, 150
Wood, as fuel, 48
  combustion properties of, 50
  ignition, effect of temperature and
    exposure time, 50
  temperatures, 51
  in construction, 207
  pyrolysis of, 66
  temperature of, 66
  surfaces, fire hazard of, 209
Wool fabrics, fire hazard of, 126